Selman Tunc Yilmaz

Exploring Single Spin Physics in Self-Assembled Quantum Dots

Selman Tunc Yilmaz

Exploring Single Spin Physics in Self-Assembled Quantum Dots

Resonance fluorescence reveals the interaction of a single spin with a single photon and with a spin bath

Südwestdeutscher Verlag für Hochschulschriften

Impressum/Imprint (nur für Deutschland/only for Germany)
Bibliografische Information der Deutschen Nationalbibliothek: Die Deutsche Nationalbibliothek verzeichnet diese Publikation in der Deutschen Nationalbibliografie; detaillierte bibliografische Daten sind im Internet über http://dnb.d-nb.de abrufbar.
Alle in diesem Buch genannten Marken und Produktnamen unterliegen warenzeichen-, marken- oder patentrechtlichem Schutz bzw. sind Warenzeichen oder eingetragene Warenzeichen der jeweiligen Inhaber. Die Wiedergabe von Marken, Produktnamen, Gebrauchsnamen, Handelsnamen, Warenbezeichnungen u.s.w. in diesem Werk berechtigt auch ohne besondere Kennzeichnung nicht zu der Annahme, dass solche Namen im Sinne der Warenzeichen- und Markenschutzgesetzgebung als frei zu betrachten wären und daher von jedermann benutzt werden dürften.

Verlag: Südwestdeutscher Verlag für Hochschulschriften GmbH & Co. KG
Heinrich-Böcking-Str. 6-8, 66121 Saarbrücken, Deutschland
Telefon +49 681 37 20 271-1, Telefax +49 681 37 20 271-0
Email: info@svh-verlag.de

Approved by: Zurich, ETH Zurich, Diss., 2011

Herstellung in Deutschland:
Schaltungsdienst Lange o.H.G., Berlin
Books on Demand GmbH, Norderstedt
Reha GmbH, Saarbrücken
Amazon Distribution GmbH, Leipzig
ISBN: 978-3-8381-3020-0

Imprint (only for USA, GB)
Bibliographic information published by the Deutsche Nationalbibliothek: The Deutsche Nationalbibliothek lists this publication in the Deutsche Nationalbibliografie; detailed bibliographic data are available in the Internet at http://dnb.d-nb.de.
Any brand names and product names mentioned in this book are subject to trademark, brand or patent protection and are trademarks or registered trademarks of their respective holders. The use of brand names, product names, common names, trade names, product descriptions etc. even without a particular marking in this works is in no way to be construed to mean that such names may be regarded as unrestricted in respect of trademark and brand protection legislation and could thus be used by anyone.

Publisher: Südwestdeutscher Verlag für Hochschulschriften GmbH & Co. KG
Heinrich-Böcking-Str. 6-8, 66121 Saarbrücken, Germany
Phone +49 681 37 20 271-1, Fax +49 681 37 20 271-0
Email: info@svh-verlag.de

Printed in the U.S.A.
Printed in the U.K. by (see last page)
ISBN: 978-3-8381-3020-0

Copyright © 2011 by the author and Südwestdeutscher Verlag für Hochschulschriften GmbH & Co. KG and licensors
All rights reserved. Saarbrücken 2011

DISS. ETH NO.

Exploring Single Spin Physics in Self-Assembled Quantum Dots Using Resonance Fluorescence

A dissertation submitted to the
SWISS FEDERAL INSTITUTE OF TECHNOLOGY ZURICH

for the degree of
Doctor of Natural Sciences

presented by
SELMAN TUNC YILMAZ

Dipl. Phys.
ETH Zürich

Prof. Ataç Imamoğlu,   examiner
Prof. Vahid Sandoghdar, co-examiner

2011

1. Foreword

A single spin confined in an optically active InGaAs quantum dot (QD) provides an exciting playground to study fundamental quantum mechanical phenomena at the single emitter level. Polarization selective spin dependent resonant optical excitations lead to life-time limited linewidths underlining the similarity with the atomic transitions. Together with the long spin life time in the ms range [1–4] and decoherence time in the μs range [5–8], a confined spin is a candidate for a quantum information processing unit [9, 10]. Experiments along these lines demonstrated high fidelity optical spin preparation [2, 11], ultrafast coherent manipulation [12, 13], n-shot [1, 14] and very recently single-shot [15] measurement of the QD spin.

On the other hand the physics of the single spin is a lot more richer because of it's in general non-trivial coupling to the solid state environment. Examples for these couplings are cotunneling and phonon-assisted spin-flip events [16] and hyperfine interaction with the $\sim 10^5$ nuclei leading to dynamic nuclear spin polarization [17–20]. In these kind of experiments the single spin serves as a probe for the solid state environment.

The first chapter of the thesis gives an overview of the sample structure and the setup. The second chapter describes various techniques of resonantly probing a single spin. After these introductory chapters I present the main results of my thesis:

First I present our results where we bring the electron spin one step closer to represent a qubit. We realize a single spin-photon interface [14] which constitutes the first step for the entanglement of distant spins which in turn is a precondition for quantum optical networks [21, 22].

In the second part I present the first measurement of the hyperfine interaction of a single QD heavy hole (HH) spin with the nuclei which was made possible with our ability to lock the degree of nuclear spin polarization to a precise value [18]. Since we observe a rather weak hyperfine interaction, this measurement supports the efforts aimed at using confined HH spins as qubits in quantum information processing.

2. Summary

In this dissertation a single spin confined in a InGaAs self-assembled quantum dot (QD) is studied by optical methods in various ways. The quantum dots are embedded in a Schottky diode heterostructure which enables control over the QD charging state. All the experiments are done on single QDs charged with one electron. We demonstrated background free resonance fluorescence measurements on these dots. An applied magnetic field lifts the degeneracy of the circularly polarized electron spin dependent transitions. The linearly polarized excitation laser addresses only one of the spin states. With our high degree of reflected laser suppression up to a factor of 10^7 and good overall photon collection efficiency (0.1%), we succeeded in realizing a single spin photon interface where the detection of a single photon initializes the QD electron to a definite spin eigenstate with a fidelity exceeding 99% and with a time uncertainty of $300\,ps$, limited only by the avalanche photodiode jitter time. High-fidelity fast spin-state initialization heralded by a single photon enables the realization of quantum information processing tasks such as nondeterministic distant spin entanglement. Given that we could suppress the measurement backaction to well below the natural spin-flip rate, realization of a quantum nondemolition measurement of a single spin could be achieved by increasing the fluorescence collection efficiency by a factor exceeding 10 using a photonic nanostructure.

In a second experiment we measure the strength and the sign of hyperfine interaction of a heavy hole with nuclear spins in single self-assembled quantum dots. Our experiments utilize the locking of a quantum dot resonance to an incident laser frequency to generate nuclear spin polarization. By monitoring the resulting Overhauser shift of optical transitions that are split either by electron or exciton Zeeman energy with respect to the locked transition using resonance fluorescence, we find that the ratio of the heavy-hole and electron hyperfine interactions is -0.09 ± 0.02 in three quantum dots. Since hyperfine interactions constitute the principal decoherence source for spin qubits, we expect our results to be important for efforts aimed at using heavy-hole spins in quantum information processing.

3. Zusammenfassung

In dieser Dissertation wird mit optischen Methoden ein einzelner Spin untersucht, der in einem selbstorganisierten InGaAs Quantenpunkt (QP) lokalisiert ist. Die QPe sind in einer Schottkydioden-Heterostruktur eingebettet, welche die Kontrolle über den Ladungszustand ermöglicht. Alle Experimente sind an einzelnen QPen durchgeführt, die mit einem Elektron geladen sind. Wir haben hintergrundfreie Resonanzfluoreszenzmessungen an diesen QPen demonstriert.

Ein Magnetfeld hebt die Entartung der zirkular polarisierten elektronspinabhängigen Übergänge auf. Das linear polarisierte Anregungslaserlicht adressiert nur einen der Spin-Zustände. Mit unserer effizienten Unterdrückung des reflektierten Laserlichtes bis zu einem Faktor 10^7 und der guten Sammeleffizienz der Photonen (0.1%) ist uns die Verwirklichung einer Spin-Photon-Schnittstelle gelungen. Die Messung eines Photons initialisiert das QP-Elektron in einem bestimmten Spin-Eigenzustand mit einer Genauigkeit über 99% und mit einer von der Lawinenfotodiode begrenzten Zeitauflösung von $300\,ps$. Schnelle Spinzustandinitialisierung mit hoher Genauigkeit, ausgelöst durch ein einziges Photon, ermöglicht die Verwirklichung von quanteninformationsverarbeitenden Prozessen wie die nichtdeterministische Verschränkung von fernen Spins. Ist die Rückwirkung der Messung deutlich unter der natürlichen Spin-Umklapprate, wäre die Verwirklichung von Nondemolitionsmessung eines Spins möglich, wenn man die Sammeleffizienz der Fluoreszenz mit Hilfe von photonischen Nanostrukturen um einen Faktor von mehr als 10 erhöhen würde.

In einem zweiten Experiment messen wir die Stärke und das Vorzeichen der Hyperfeinwechselwirkung eines schweren Lochs mit Kernspins in einzelnen selbstorganisierten QPen. Um eine Kernspinpolarization zu erzeugen, nutzen wir die Fixierung der QP-Resonanz auf die Frequenz des einfallenden Laserlichtes. Durch die Beobachtung der daraus resultierenden Overhauser-Verschiebung der optischen Übergänge, die in Bezug auf den fixierten Übergang entweder durch Elektronen- oder Exzitonen-Zeeman-Energie gespalten sind, finden wir, mit Hilfe der Resonanzfluoreszenz, dass das Verhältnis der Hyperfeinwechselwirkungen des schweren Loches und des Elektrons in drei QPen -0.09 ± 0.02 ist. Da die Hyperfeinwechselwirkung die wichtigste Dekohärenzquelle für Spin-Qubits bildet, erwarten wir, dass unsere Ergebnisse wichtig für die Bemühungen sind, den Spin des schweren Lochs in der Quanteninformationsverarbeitung zu verwenden.

Contents

Titel	**a**
1. Foreword	**c**
2. Summary	**e**
3. Zusammenfassung	**g**
Contents	**i**
4. Introduction	**1**
4.1. Charge Tunable Self-assembled InGaAs Quantum Dots	1
4.2. Cryogenic Confocal Microscopy	4
4.3. Photon collection efficiency	6
5. Resonantly probing a single electron spin	**9**
5.1. Energy levels	9
5.2. Optically driven 2-level quantum emitter	11
5.3. Optical cooling of a single electron spin	15
5.4. Signal-to-noise ratio analysis for time-resolved electron spin measurement	18
5.5. Measurement in other polarization basis	20
6. Quantum-dot-spin single-photon interface	**25**
6.1. Introduction	25
6.2. Background-free resonance fluorescence measurement	25
6.3. Study of electron spin dynamics using photon correlation	28
6.4. Calculation of the 2nd order photon correlation $G_2(\tau)$	29
7. Measurement of heavy-hole hyperfine interaction in InGaAs quantum dots using resonance fluorescence	**33**
7.1. Introduction	33
7.2. Measurements	34
7.3. Nuclear spin polarization by driving the diagonal transition	39
7.4. Conclusions	41
8. Outlook	**43**
8.1. Distant spin-spin entanglement	43
8.2. Quantum nondemolition (QND) measurement	44
8.3. Electron spin resonance	45
8.4. Measurement of in-plane hole hyperfine interaction	45

8.5. Study of heavy-hole light-hole mixing 45

A. Bibliography   I

Acknowledgment   VII

List of Figures   IX

4. Introduction

4.1. Charge Tunable Self-assembled InGaAs Quantum Dots

The self-assembled InGaAs quantum dots (QDs) are islands of InGaAs buried in GaAs (Fig. 4.1, left). In order to create these islands few mono layers of InAs are deposited in an MBE machine on the epitaxial GaAs. Due to the lattice constant mismatch between the InAs and GaAs, when a critical height is reached these islands form naturally in order to reduce the stress ([23–26] and the references therein). The smaller band gap of InAs enables confinement of the electrons and holes within these islands in all three dimensions, with respective ground state confinement energies of 130 meV and 80 meV [27]. The QD boundary is sharp in the growth direction and rather smooth in the lateral directions. Together with the typical dot dimensions of $\sim 5nm$ in the vertical and $\sim 20\,nm$ in the lateral, the vertical confinement is much more stronger than the lateral confinement. The lateral confinement can be approximated by a harmonic potential leading to equi-spaced single particle levels, with the electron and hole quantization energies of $\hbar\omega_e = 50\,meV$ and $\hbar\omega_h = 25\,meV$ [28]. Due to the large level spacing of the vertical confinement all the bound states correspond to the ground state of the vertical confinement.

Fig. 4.1 shows in more detail the different layers of the sample in the growth direction. The first epitaxial layer is a GaAs buffer layer grown on top of GaAs substrate. The next layer is a highly Si-doped GaAs which serves as electron reservoir. The neighboring 35 nm GaAs is a barrier preventing the electrons from tunneling between the QD and the doped layer. Although the QDs are formed by self-assembly, a continuous 2D InAs layer remains. This so called wetting layer is basically a quantum well. QDs are little bumps on the wetting layer. On top of the QD/wetting layer, 12 nm thick GaAs spacer layer protects the dot from the charge fluctuations intrinsic to the GaAs/AlAs supperlattice (the blocking barrier). The spacer layer should be thin enough such that 2D hole levels in the triangular potential formed by the spacer layer and the blocking barrier are detuned from the QD hole levels. The blocking barrier prevents the electron tunneling from the top gate and reduces the current in the sample drastically. The capping layer protects the structure from the fluctuating electrical environment of the surface.

A 4 nm of a Ti gate deposited with physical vapor deposition on the capping layer creates a Shottky barrier with a built-in voltage V_g^0. The Fermi level of the structure is pinned at the conduction band of the doped layer. By tuning the gate voltage the position of the quantized QD levels can be changed with respect to the Fermi level, making the deterministic charging of the QD possible [29]. The band digram together with the dot levels goes down (up) with respect to the Fermi level when a positive (negative) voltage on the Shottky gate is applied. For a given gate

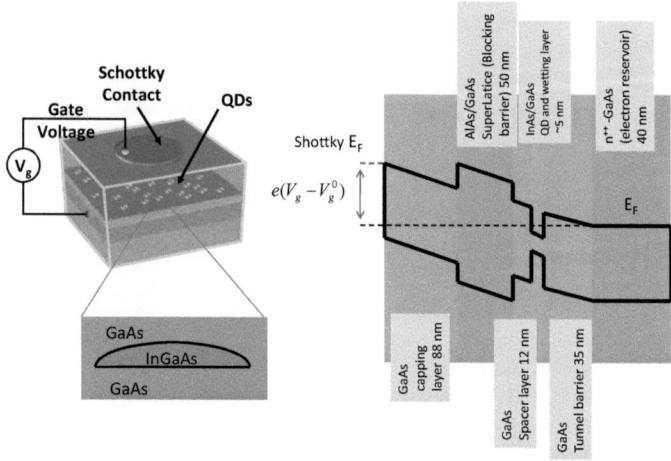

Figure 4.1.: Sketch of the sample structure (left) together with the band diagram along the growth direction (right). V_g^0 is the built-in voltage due to Shottky barrier and V_g is the applied voltage.

voltage, if the lowest quantized conduction band QD level is above the Fermi level, no electron can tunnel into the QD and the ground state is an empty QD, designated as X^0. By increasing the gate voltage the lowest QD level comes into resonance with the Fermi level and an electron from the doped layer tunnels into the dot. The dot is now single electron charged and designated as X^-. Although the spin degree of freedom allows a second electron to tunnel into this state, due to the Coulomb repulsion between the electrons the two-electron state has a higher energy than the Fermi level and the dot remains singly charged. This is called Coulomb blockade [28]. The charging energy is $\sim 20\,meV$. By increasing the gate voltage the dot can be sequentially filled with electrons occupying the higher lying levels of the lateral confinement potential.

The QD charge control can nicely be seen when μ-photoluminescence (μ-PL) spectrum is plotted as a function of gate voltage (Fig. 4.2). Here the dot is excited with a 780 nm laser which creates free electron hole (e-h) pairs in GaAs and in the wetting layer. The excitation mainly provides free holes to be captured by the QD, whereas electrons are provided mainly by the doped layer. Due to carrier-phonon and carrier-carrier scattering they diffuse around the sample and some of them get trapped in the dot potential within $35\,ps$ [30], and relax within $\sim 40\,ps$ to the dot ground state [31]. When the electron and hole are on the respective ground states e-h recombination occurs within $1\,ns$ [32] and a photon is emitted. Due to Coulomb interactions between the carriers in the QD, the photon energy depends highly on the stable charging state of the QD before an e-h pair is captured. In fig. 4.2 different charging states are seen as discrete plateaus. In the figure two plateaus at the low gate voltage values are assigned to the positively charged QD states such as

4.1. Charge Tunable Self-assembled InGaAs Quantum Dots

the single hole charged X^+ and double hole charged X^{++}. At these gate voltages electrons quickly tunnel out from the QD but the holes can stay trapped. In the case of X^+ the Coulomb interaction of the two holes with the electron can bind an electron to the dot. When an e-h pair recombines a single hole is left in the dot. X^{++} corresponds to the scenario when the three holes bind the electron before the e-h recombination. In the plot different plateaus are overlapping for certain gate voltage regions. These are because of the finite tunneling rates of the holes and the electrons and do not necessarily correspond to stable QD ground states. There are no overlapping regions when the QD levels are probed resonantly, creating the carriers only within the dot potential.

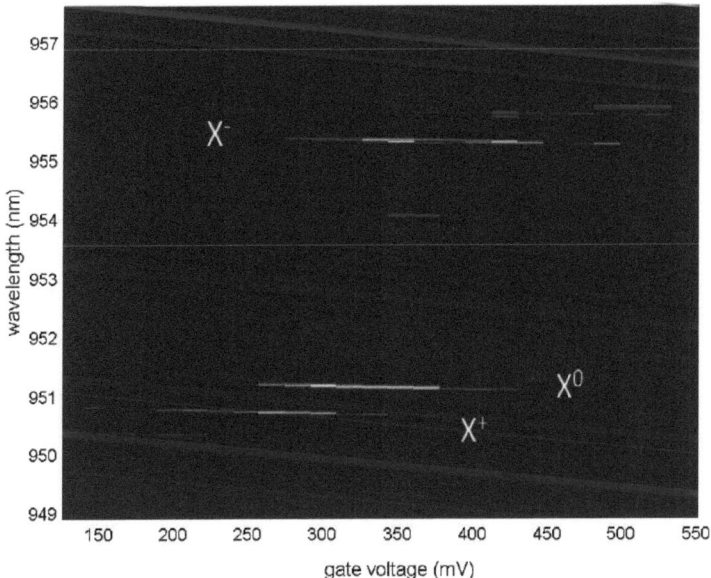

Figure 4.2.: μ-photoluminescence spectrum as a function of gate voltage.

Another observable feature is that the plateaus are tilted such that the energy increases with increasing gate voltage. This is the DC Stark effect and the change in energy with the electric field is described by the formula [33]

$$\Delta E = -d_{perm}\epsilon + \beta_{qd}\epsilon^2 \tag{4.1}$$

where d_{perm} is the permanent dipole moment of the e-h pair, β_{qd} is the polarizability and ϵ is the electric field at the dot location.

Fig. 4.3 shows the magnetic field dependence of the resonances of X^- probed by a resonant laser at a fixed gate voltage. The upper (lower) curve corresponds to the resonance when the resident QD electron is in the spin-up (-down) state. The change in energy with magnetic field is given by [34]

$$\Delta E = \pm \frac{1}{2}\mu_B(|g_h| + |g_e|)B + \alpha B^2 \qquad (4.2)$$

where μ_B is the Bohr magneton, g_e (g_h) is the electron(hole) g-factor, α is a constant for the diamagnetic shift and $+(-)$ sign is for spin-up(down) electron state. In the figure the fit parameters are $\frac{1}{2}\mu_B(|g_h|+|g_e|) = 54\,\mu eV/T$ ($|g_h|+|g_e| = 1.9$) and $\alpha = 9\mu eV/T^2$. Electron g-factor does not vary from to dot and is $g_e = -0.6$ whereas the hole g-factor exhibits a large deviation. In our experiments g_h was in the range of $0.9 - 1.4$. The diamagnetic shift originates from the modification of the ground and excited state wavefunctions under the magnetic field. The diamagnetic shift scales as B^2 in the limit where electron (hole) cyclotron frequency $\omega_{e(h)}^{cr} = \frac{eB}{m_{e(h)}}$ is much smaller than the single particle frequency $\Omega_{e(h)}$ of the parabolic lateral confining potential $V_{e(h)} = \frac{m_{e(h)}\Omega_{e(h)}r^2}{2}$ [35, 36].

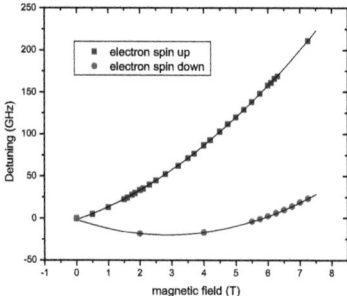

Figure 4.3.: Magnetic field dependence of the X^- resonances for the electron spin-up and spin-down transitions.

4.2. Cryogenic Confocal Microscopy

In order to access the ground state transitions of the QD the sample is in a liquid helium bath cryostat at 4 K. With the confocal microscope a single quantum dot can be addressed (Fig. 4.4). The bottom part of the microscope, as depicted in fig. 4.4, is housed in a vacuum tight steel tube. After evacuating the tube it is filled with 20 mbar of He for thermal contact of the sample. The microscope has three arms coupled to fibers. These can be used either for excitation or for photon collection. The sample can be moved with an accuracy of $\approx 100\,nm$ in all 3 dimensions with the attocube positioners in order to bring a single QD to the focal spot. All of the fibers are single mode fibers. The polarization of the excitation lasers can be adjusted with $\lambda/2$ plates and polarizers on the microscope arms. The objective is achromatic, the z-position of the sample is adjusted such that the QD is in focus for it's characteristic emission wavelength.

In the μ-PL measurements, the dot is excited with a quasi-focused laser at 780 nm. The PL photons are collected with one arm of the microscope and sent with the fiber

4.2. Cryogenic Confocal Microscopy

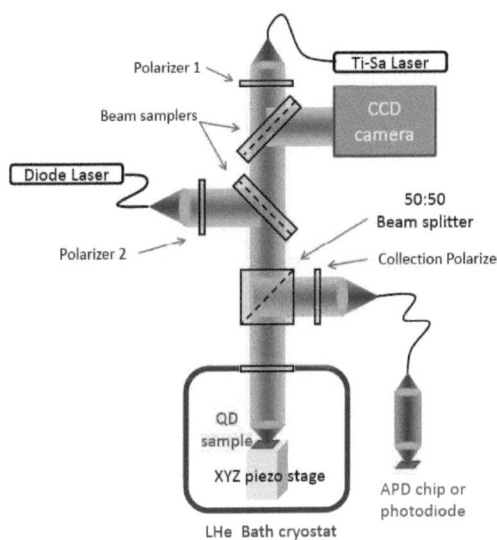

Figure 4.4.: Sketch of the confocal microscope immersed into the liquid helium bath cryostat.

to the spectrometer. For absorption measurements in transmission a PIN photodiode is placed directly below the sample. A preamplifier outside of the cryostat converts the photo current to voltage.

For resonance fluorescence (RF) measurements the sample is excited with a linearly polarized light. The RF photons are collected with the lowest microscope arm and sent to an avalanche photodiode (APD) in the Geiger mode. The collection polarizer is oriented perpendicularly to the laser polarization. With this cross-polarization scheme the reflected laser photons can be suppressed up to a factor of 10^7. This high level of suppression is only possible by careful alignment of the microscope. Good suppression is achieved only when the excitation laser polarization is along the axis of the 50/50 beam cube and the beam samplers (p-or s-polarized) , otherwise these elements create elliptical polarization. The beam paths have to be perpendicular to the optical elements in the microscope, including especially the polarizers. The polarization suppression is the most sensitive aspect of the measurements to alignment and hence to the mechanical stability of the microscope. The use of the single mode fiber in the collection path is crucial for polarization suppression. We think that upon reflection from the sample and after passing through many optical elements the laser beam is no longer in pure Gaussian mode and the other additional modes have polarizations deviating from the linear. The single mode fiber filters these other modes and hence the mixture from the non-linear polarization components.

The two beam samplers transmit most of the light. They are highly polarization sensitive. Since they displace the beam, they are placed with opposite orientation in order to compensate for the displacement. The CCD image of the focal spot helps for alignment. Not shown in the figure are the opto-mechanical elements on the three arms for aligning the optical paths with respect to the optical axis defined by the objective. On various parts of the microscope phodiodes are placed for laser power stabilization.

In order to align a microscope arm, a 50/50 fiber beam splitter is connected to the arm and the reflected light from the sample, being collected by one port of the fiber beam splitter, is maximized by moving the attocubes and the opto-meachanical elements of the respective arm. A second arm is aligned with respect to this first arm by sending light from the second arm and maximizing the reflected signal in the first arm. Only the opto-mechanical elements of the second arm are aligned in this case.

4.3. Photon collection efficiency

The main limitation for the collection efficiency is the total internal reflection taking place when the emitted QD photons within GaAs with the high refractive index of $n_{GaAs} = 3.5$ cross the GaAs-air boundary. Only the photons within the cone bounded by the critical angle θ_c for the total internal reflection can cross the boundary. The critical angle is given by

$$\theta_c = \sin^{-1}(\frac{n_{air}}{n_{GaAs}}) = 16.6^0 \qquad (4.3)$$

The combination of a high NA ($NA_{objective} = 0.55$) aspherical objective together with a hemispherical cubic zirconia ($n_{zirconia} = 2.13$) solid immersion lens (SIL) is used (Fig. 4.5). Together the NA of the system is $n_{zirconia}NA_{objective} = 1.18$, which means that all the light falling on the planar zirconia interface from the air side can be collected by the objective, whereas some of the light falling on the boundary from the zirconia side is total internally reflected. It is assumed that SIL does not make an optical contact with the GaAs.

To address the first issue a very thin vacuum grease layer ($n_{grease} \approx 1.5$) is incorporated between the SIL and the GaAs (Fig. 4.5). Making optical contact with GaAs, its effect is to increase the critical angle θ_c to θ_c' as

$$\sin\theta_c' = \sin\theta_c \frac{n_{grease}}{n_{air}} \qquad (4.4)$$

with θ_c as defined in 4.3.
Fromm Snell's Law the equations at the two interfaces are

$$n_{zirconia}\sin\theta_1 = n_{grease}\sin\theta_2 \qquad (4.5)$$
$$n_{grease}\sin\theta_2 = n_{GaAs}\sin\theta_3 \qquad (4.6)$$

Since for the maximum of θ_1^{max} with $\sin\theta_1^{max} = 0.55$, from 4.6 follows

4.3. Photon collection efficiency

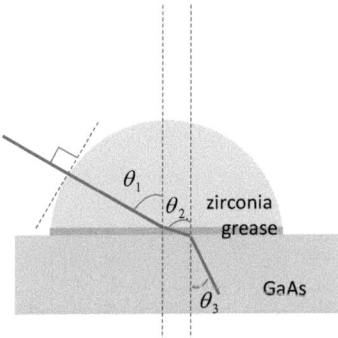

Figure 4.5.: The path of the light ray from a beam focused on the sample.

$$\sin\theta_2^{max} = \frac{n_{zirconia}}{n_{grease}}\sin\theta_1^{max} = 0.78 < 1 \quad (4.7)$$

which means that the objective cannot collect all the photons falling on the zirconia-grease interface from the grease side. 4.7 allows to write

$$n_{zirconia}\sin\theta_1^{max} = n_{grease}\sin\theta_2^{max} = n_{GaAs}\sin\theta_3^{max} \quad (4.8)$$

$$\Rightarrow \theta_3^{max} = \sin^{-1}(\frac{n_{zirconia}}{n_{GaAs}}0.55) = 19.5^0 \text{ (0.34 in radians)} \quad (4.9)$$

Without a grease layer $\theta_3^{max} = \theta_c = 16.6^0$ since in this case all the light falling in the flat surface of the SIL from the air side can be collected by the objective, as noted earlier. In order to calculate the collection efficiency the procedure in [37] is followed. The intensity distributions of the dipole emission for the s and p polarized light are written as

$$I_s(\theta,\phi) = \frac{3}{8\pi}[1-\sin^2(\theta)\cos^2(\phi)]\sin^2(\phi) \quad (4.10)$$

$$I_p(\theta,\phi) = \frac{3}{8\pi}[1-\sin^2(\theta)\cos^2(\phi)]\cos^2(\phi) \quad (4.11)$$

The distributions are normalized such that $\int_{\phi=0}^{\phi=2\pi}\int_{\theta=0}^{\theta=\phi}d\theta d\phi\sin(\theta)(I_p(\theta,\phi) + I_s(\theta,\phi)) = 1$. The collection efficiency is then given as

$$\int_{\phi=0}^{\phi=2\pi}\int_{\theta=0}^{\theta=0.34}d\theta d\phi\sin(\theta)(I_p(\theta,\phi)(1-R_p(\theta,\phi)) + I_s(\theta,\phi)(1-R_s(\theta,\phi)))$$

$$= 0.037 \quad (4.12)$$

where R_s (R_p) is the intensity reflectivity at the GaAs/grease interface for the s(p)-polarized light. As a result of the smaller refractive index difference the reflection at the grease/zirconia interface is negligible. Taking into account the various losses along the optical path such as the 50/50 beam splitter (0.5), the coupling into the single mode fiber (0.3), APD quantum efficiency (0.25), we get for the overall collection efficiency

$$0.5 \times 0.3 \times 0.25 \times 0.037 \approx 1.4 \times 10^{-3}. \tag{4.13}$$

This is very close to the value we estimate from the RF counts from a strongly excited QD. We collect 5×10^5 photons per second from a QD emitting 0.5×10^9 photons per second, assuming a spontaneous emission rate of $\hbar\gamma_{sp} = 10^9\,s^{-1}$, resulting in the collection efficiency of 10^{-3}. The factor of 1.4 discrepancy can be explained by the uncertainty inthe measurement of Γ_{sp} and by the 4 nm titanium Shottky gate which has not been taken into account in the calculation. From transmission measurements, it is known that it reflects 40% of the laser power focussed onto the sample. Therefore it is expected that the reflection of the RF photons is also enhanced due to the metallic layer.

5. Resonantly probing a single electron spin

In this chapter the interaction of a quasi-resonant laser with a single electron charged QD is discussed. Considering the QD as a 2-level quantum emitter in the focus of a Gaussian beam the resonance fluorescence and interference signals are calculated. After describing the phenomenon of optical cooling of the electron spin, in the presence of measurement induced spin cooling, a procedure to calculate the signal-to-noise ratio for a single shot spin measurement is given. Finally a generalized measurement method for quasi-resonant excitation is described.

5.1. Energy levels

Fig. 5.1 illustrates the relevant energy levels of a single electron charged QD under external magnetic field in the sample growth direction (z-direction, Faraday geometry). In the ground state a single electron resides in the QD, either in spin-up $|\uparrow\rangle$ or spin-down $|\downarrow\rangle$ eigenstate. The two optically excited (trion) states $|\uparrow\downarrow\Uparrow\rangle$ and $|\uparrow\downarrow\Downarrow\rangle$ are formed by two electrons in the singlet state and a heavy hole, either in the (pseudo) spin up $|\Uparrow\rangle$ or down $|\Downarrow\rangle$ state. The two ground and excited states are split by the Zeeman energies $|g_e|\mu_B B$ and $|g_h|\mu_B B$. A linearly polarized laser with its k-vector along the z-direction excites the QD. The optical selection rules in the Faraday geometry are such that its right (left)-hand circularly polarized component drives the transition $|\uparrow\rangle \leftrightarrow |\uparrow\downarrow\Uparrow\rangle$ ($|\downarrow\rangle \leftrightarrow |\downarrow\uparrow\Downarrow\rangle$) with the Rabi frequency Ω_R (Ω_L). In fig. 5.1 the interactions of the level $|4\rangle$ are omitted for clarity. Spontaneous emission from the trion state occurs with rate $\Gamma \sim 10^9 s^{-1}$ and is circularly polarized. The selection rules are relaxed by the ground state mixing due to hyperfine interaction of the electron with the nuclei and by the heavy hole(hh) - light hole(lh) mixing, resulting in weak diagonal transitions with rate γ. For fields larger than 600 mT the ground state mixing is negligible and the heavy-light hole mixing is the main mechanism for the diagonal decay [1].

The Hamiltonian in the Schrödinger picture is given by

$$H = \frac{\hbar\Omega_R}{2}(e^{i\omega_L t}|1\rangle\langle 3| + hc) + \frac{\hbar\Omega_L}{2}(e^{i\omega_L t}|2\rangle\langle 4| + hc) + \frac{\mu_B}{2}g_e\hat{\sigma}_{z,e}B + \frac{\mu_B}{2}g_h\hat{\sigma}_{z,h}B + \hbar\omega_0(|3\rangle\langle 3| + |4\rangle\langle 4|) + \hbar\Omega_N(|1\rangle\langle 2| + |2\rangle\langle 1|) \quad (5.1)$$

where $\hbar\omega_L$ is the laser energy and $\hbar\omega_0$ is the mean of the $|1\rangle \leftrightarrow |3\rangle$ and $|2\rangle \leftrightarrow |4\rangle$ transition energies, $\hbar\Omega_N$ is the ground state coupling due to hyperfine interaction. $\hat{\sigma}_{z,e}$ ($\hat{\sigma}_{z,h}$) is the z-component of the Pauli vector for the ground (trion)-states defined as

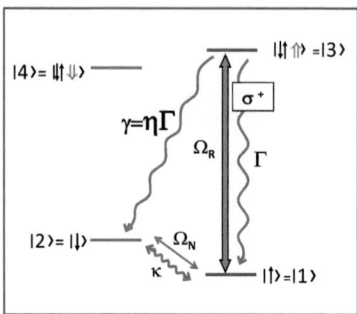

Figure 5.1.: Level diagram describing the singly charged QD in the presence of an external magnetic field in the z-direction. For clarity the transitions involving the level $|\downarrow\uparrow\Downarrow\rangle$ are not shown. Shown are the coupling of the blue vertical transition to the σ^+-polarized light field with Rabi frequency Ω_R, the spontaneous emission rate Γ, the weak diagonal decay rate $\gamma = \eta\Gamma$ with $\eta \ll 1$ being the branching ratio, the hyperfine interaction induced electron spin state mixing with the strength $\hbar\Omega_N$ and the electron spin flip rate κ as a result of the cotunneling processes.

$$\hat{\sigma}_{z,e} = |1\rangle\langle 1| - |2\rangle\langle 2| \quad (5.2)$$
$$\hat{\sigma}_{z,h} = |3\rangle\langle 3| - |4\rangle\langle 4| \quad (5.3)$$

For a linearly polarized excitation laser along **x**-direction with the electric field vector at the focus $(x = 0, y = 0, z = 0)$

$$\mathbf{E}_L = \mathbf{E}_0 \cos(\omega_L t)\,\hat{\mathbf{x}} \quad (5.4)$$

The Rabi frequencies are given by

$$\Omega_R^L = e\mathbf{E}_0\mathbf{D}|\langle \mathbf{x}|\sigma^+_-\rangle|^2/\hbar \quad (5.5)$$

where **D** is the transition dipole moment. When ω_L is close to the transition frequency $|1\rangle \leftrightarrow |3\rangle$, the coupling to the state $|4\rangle$ can be safely neglected for fields $B > 0.2\,T$.

In order to eliminate the time dependence of the Hamiltonian one can go to the rotating frame transforming the basis states as $|\tilde{3}\rangle = |3\rangle e^{-i\omega_L t}$ and $|\tilde{4}\rangle = |4\rangle e^{-i\omega_L t}$. In this frame the time dependence of the density matrix $\hat{\rho}(t)$ is given by [16, 38]

$$\dot{\hat{\rho}}(t) = \frac{1}{i\hbar}[H, \hat{\rho}(t)] + L \quad (5.6)$$

5.2. Optically driven 2-level quantum emitter

where the Liouvillian L is given by

$$\begin{aligned}L =\ & \frac{\Gamma}{2}(2\hat{\sigma}_{13}\hat{\rho}\hat{\sigma}_{31} - \hat{\sigma}_{33}\hat{\rho} - \hat{\rho}\hat{\sigma}_{33}) + \frac{\Gamma}{2}(2\hat{\sigma}_{24}\hat{\rho}\hat{\sigma}_{42} - \hat{\sigma}_{44}\hat{\rho} - \hat{\rho}\hat{\sigma}_{44}) +\\ & (1+\bar{n})\frac{\kappa}{2}(2\hat{\sigma}_{12}\hat{\rho}\hat{\sigma}_{21} - \hat{\sigma}_{22}\hat{\rho} - \hat{\rho}\hat{\sigma}_{22}) + \bar{n}\frac{\kappa}{2}(2\hat{\sigma}_{21}\hat{\rho}\hat{\sigma}_{12} - \hat{\sigma}_{11}\hat{\rho} - \hat{\rho}\hat{\sigma}_{11}) +\\ & \frac{\eta\Gamma}{2}(2\hat{\sigma}_{23}\hat{\rho}\hat{\sigma}_{32} - \hat{\sigma}_{33}\hat{\rho} - \hat{\rho}\hat{\sigma}_{33}) + \frac{\eta\Gamma}{2}(2\hat{\sigma}_{14}\hat{\rho}\hat{\sigma}_{41} - \hat{\sigma}_{44}\hat{\rho} - \hat{\rho}\hat{\sigma}_{44})\end{aligned} \quad (5.7)$$

where $\hat{\sigma}_{ij} = |\tilde{i}\rangle\langle\tilde{j}|$, Γ is the spontaneous emission rate due to coupling the vacuum photon states and $\eta = \frac{\gamma}{\Gamma}$ is the branching ratio. Cotunneling processes with the electrons in the n-doped layer result in the spin flip rates $(1+\bar{n})\kappa$ and $\bar{n}\kappa$ [16, 39], with $\bar{n} = \frac{1}{\exp(g_e\mu_B B/k_B T)-1}$ being the Boltzman factor.

5.2. Optically driven 2-level quantum emitter

In order to calculate the signal from the resonantly driven QD, first an optically driven 2-level system (levels $|1\rangle$ and $|3\rangle$) with the transition energy $\hbar\omega_{13}$ will be studied and the level $|2\rangle$ will be incorporated later.

A Gaussian beam is focused onto the QD with the electric field at the focal plane written as

$$\hat{\mathbf{E}}_L = \hat{\mathbf{E}}_L^+ + \hat{\mathbf{E}}_L^- \quad (5.8)$$

$$\mathbf{E}_L^\pm(x,y) = \pm i\frac{A_0}{2}e^{-\frac{x^2+y^2}{v_0^2}}e^{\mp i\omega_L t}\hat{\mathbf{x}} \quad (5.9)$$

The geometry of the experiment is sketched in fig. 5.2.

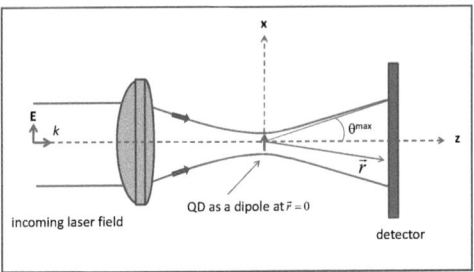

Figure 5.2.: Geometry of the QD excitation. A linearly polarized Gaussian laser beam is focused on the QD considered as a dipole along the x-direction. A detector in the far field measures the composite field over the solid angle bounded by θ^{max}.

The laser is linearly polarized along the $\hat{\mathbf{x}}$-direction and the dipole moment of the 2-level emitter is taken to be copolarized, i.e. $\mathbf{D} = D\hat{\mathbf{x}}$. The emission of a circularly polarized dipole is the superposition of two linearly polarized orthogonal dipoles with a phase difference of $\frac{\pi}{2}$. Starting with the expression in [40] the electric field operator of the emitter in the far field is written as

$$\hat{\mathbf{E}}_{QD} = \hat{\mathbf{E}}_{QD}^+ + \hat{\mathbf{E}}_{QD}^- \qquad (5.10)$$

$$\hat{\mathbf{E}}_{QD}^+(\theta,\phi,r) = -i\frac{e\omega_{13}^2 D}{4\pi\epsilon_0 c^2 r}\begin{pmatrix} 1-\sin^2\theta\cos^2\phi \\ -\sin^2\theta\cos\phi\sin\phi \\ \sin\theta\cos\theta\cos\phi \end{pmatrix}\hat{\pi}(t-\frac{r}{c}) \qquad (5.11)$$

where $\hat{\mathbf{E}}_{QD}^-$ is the hermitian conjugate of $\hat{\mathbf{E}}_{QD}^+$ and $\hat{\pi}(t) = |1\rangle\langle 3|$ is the lowering operator in the Heisenberg picture with $\langle\hat{\pi}(t)\rangle = \rho_{31}(t)e^{-i\omega_L t}$. The steady state solution is

$$\rho_{31} = -\frac{\Omega}{2}\frac{-\Delta\omega + i\frac{\Gamma_{lw}}{2}}{\Delta\omega^2 + \frac{\Gamma_{lw}^2}{4} + \frac{\Gamma_{lw}}{2\Gamma}\Omega^2} \qquad (5.12)$$

$$\Gamma = \frac{e^2\omega_{13}^3 D^2}{3\pi\epsilon_0 c^3 \hbar} \qquad (5.13)$$

where Γ is the spontaneous emission rate, $\Gamma_l w$ is the total line width with $\Gamma = \Gamma_{lw}$ when there is no dephasing, $\Delta\omega = \omega_L - \omega_{13}$ and $\Omega = eA_0 D/\hbar$. A detector positioned in the far field below the sample detects the interference signal of the dot field and laser field.

Here the laser field is given by [41]

$$\mathbf{E_L}^\pm(\mathbf{r},t) = b_0\frac{z}{r^2}e^{\pm i(\mathbf{kr}-\omega_L t)}e^{-\frac{1}{c_0^2}\frac{x^2+y^2}{r^2}}(z\hat{\mathbf{x}} - x\hat{\mathbf{z}}) \qquad (5.14)$$

$$b_0 = \frac{A_0\pi\eta_0^2}{2\lambda} \qquad (5.15)$$

$$c_0^2 = \frac{4}{k^2\eta_0^2} \qquad (5.16)$$

where $c_0 = \theta_{max}$ is the angular spread of the Gaussian beam. Note the Guoy phase shift [42] of $-\frac{\pi}{2}$ in the far field with respect to the field at the focus. The total detected power is

$$P = \int_{det} I_{total}(\mathbf{r})\hat{\mathbf{u}}_c(\mathbf{r})d\mathbf{s} \qquad (5.17)$$

$$I_{total}(\mathbf{r}) = 2\epsilon_0 c(\mathbf{E}_L^- + \mathbf{E}_{QD}^-)(\mathbf{E}_L^+ + \mathbf{E}_{QD}^+) \qquad (5.18)$$

$$= 2\epsilon_0 c\langle\hat{\mathbf{E}}_{QD}^-\hat{\mathbf{E}}_{QD}^+\rangle + 2\epsilon_0 c(\mathbf{E}_L^-\mathbf{E}_L^+) + 4\epsilon_0 c Re\{\mathbf{E}_L^+\mathbf{E}_{QD}^-\} \qquad (5.19)$$

The integrals are over the detector area, $\hat{\mathbf{u}}_c(\mathbf{r})$ is the unit propagation vector of the composite field at position $\mathbf{r}$, $d\mathbf{s}$ is the surface element on the detector. The last term is the interference term. The ratio of the coherently scattered QD photons to the total scattered photons is given by

$$\frac{I_{coh}^{QD}}{I_{sc}^{QD}} = \frac{\langle\hat{E}^-\rangle\langle\hat{E}^+\rangle}{\langle\hat{E}^-\hat{E}^+\rangle} = \frac{(\Delta\omega^2 + \frac{\Gamma_{lw}^2}{4})\frac{\Gamma}{\Gamma_{lw}}}{\Delta\omega^2 + \frac{\Gamma_{lw}^2}{4} + \frac{\Gamma_{lw}}{2\Gamma}\Omega^2} \qquad (5.20)$$

5.2. Optically driven 2-level quantum emitter

Only the coherent part of the QD emission contributes to interference. Note that due to the Guoy phase shift, the laser interferes only with the part of the QD field originating from the imaginary part of ρ_{31} and it is a destructive interference. The total power scattered by the dot is given by

$$P_{scat} = r^2 \int d\Omega 2\epsilon_0 c \langle \hat{\mathbf{E}}_{QD}^- \hat{\mathbf{E}}_{QD}^+ \rangle$$
$$= \frac{e^2 \omega_{13}^4 D^2}{8\pi^2 \epsilon_0 c^3} \rho_{33} \int_0^{2\pi} \int_0^{\pi/2} \sin^3\theta d\theta d\phi = \Gamma \hbar \omega_{13} \rho_{33} \tag{5.21}$$

with

$$\rho_{33} = \langle \hat{\pi}^+ \hat{\pi} \rangle = \frac{\frac{1}{2}\frac{\Gamma_{lw}}{2\Gamma}\Omega^2}{\Delta\omega^2 + \frac{\Gamma_{lw}^2}{4} + \frac{\Gamma_{lw}}{2\Gamma}\Omega^2} \tag{5.22}$$

is the trion population in steady state. The total laser power hitting the sample is

$$P_L = \int_\infty I_L \hat{\mathbf{u}}_L ds = 2\epsilon_0 c \int_\infty (\mathbf{E}_L^- \mathbf{E}_L^+) \hat{\mathbf{u}}_L ds \tag{5.23}$$
$$= \frac{\epsilon_0 c A^2 \pi \nu_0^2}{4} = A_L I_0 \tag{5.24}$$

where

$$A_L = \frac{\pi \nu_0^2}{2} \tag{5.25}$$

is the laser are with $A_L \approx 1.13\, FWHM^2$ and

$$I_0 = \frac{1}{2}\epsilon_0 c A_0^2 \tag{5.26}$$

is the intensity at the QD location. Therefore, using the definitions of Γ, ρ_{33} and Ω the total scattered power by the QD can be written as

$$P_{scat} = \Gamma \hbar \omega_{13} \rho_{33} = P_L \frac{\sigma}{A_L} \tag{5.27}$$

$$\sigma = \sigma_0 \frac{\frac{\Gamma_{lw}^2}{4}}{\Delta\omega^2 + \frac{\Gamma_{lw}^2}{4} + \frac{\Gamma_{lw}}{2\Gamma}\Omega^2} \tag{5.28}$$

where

$$\sigma_0 = \frac{3(\lambda_L/n)^2}{2\pi}\frac{\Gamma}{\Gamma_{lw}} \tag{5.29}$$

is the on resonance low power cross-section of a two-level system, n is the refractive index of the medium. In this case the medium is GaAs with $n_{GaAs} = 3.5$. Making use of the the narrowness of the line width i.e $\Delta\omega \ll \omega_{13}$, ω_L, λ_L rather than λ_{13} is used in the expression for σ_0. Conservation of energy dictates that the calculated

P_{scat} is the decreased laser power in transmission due to destructive interference. Denoting it as the absorption signal P_{abs}, the following equality holds

$$P_{scat} = P_L \frac{\sigma}{A_L} = -\int_\infty 4\epsilon_0 c Re\{\mathbf{E}_L^- \mathbf{E}_{QD}^+\} ds = P_{abs} \quad (5.30)$$

The last equality has been explicitly calculated in [23] in the limit of paraxial approximation. In practice, when the collection efficiency is taken into account the detected signals are $P_{scat}^{det} = \epsilon_{RF} \frac{\sigma}{A_L} P_L$ and $P_{abs}^{det} = \epsilon_{abs} \frac{\sigma}{A_L} P_L$ with $\epsilon_{RF}, \epsilon_{abs} < 1$. Both of these detected signals have the same dependence on P_L and $\Delta\omega$. With P_L^{det} denoting the detected laser power below the sample, P_{sat} can be defined as the detected laser power at which $\sigma = \sigma_0/2$ for $\Delta\omega = 0$

$$\frac{\Gamma_{lw}}{2\Gamma}\Omega^2 = \frac{P_L^{det}}{P_{sat}} \frac{\Gamma_{lw}^2}{4} \quad (5.31)$$

$$\Rightarrow \sigma = \sigma_0 \frac{\frac{\Gamma_{lw}^2}{4}}{\Delta\omega^2 + \frac{\Gamma_{lw}^2}{4} + \frac{P_L^{det}}{P_{sat}} \frac{\Gamma_{lw}^2}{4}} \quad (5.32)$$

For both P_{scat}^{det} and P_{abs}^{det}, the FWHM of the ω_L dependence is given by

$$FWHM = \Gamma_{lw} \sqrt{1 + \frac{P_L^{det}}{P_{sat}}} \quad (5.33)$$

The contrasts at $\Delta\omega = 0$ have the same power dependence showing saturation behavior:

$$\frac{P_{scat}^{det}}{P_L^{det}}, \frac{P_{abs}^{det}}{P_L^{det}} \propto (1 + \frac{P_L^{det}}{P_{sat}})^{-1} \quad (5.34)$$

A typical resonance fluorescence (RF) gate voltage scan with a laser intensity around QD saturation together with the power dependence of the contrast at $\Delta\omega = 0$ and FWHM of the linewidth $\Delta\omega$ are shown in fig. 5.3. The magnetic field $B = 0\,T$ and the laser is linearly polarized. Depending on the electron spin state at a given time, the laser couples either to the $|1\rangle \leftrightarrow |3\rangle$ or to the $|2\rangle \leftrightarrow |4\rangle$ transition and the RF photons are coming from the respective transition. The laser power is measured in units of the photo current from a photo diode positioned below the sample. In order to get the power dependences individual scans are fitted with Lorentzian curves and the linewidths and the contrasts are extracted. In order to extract the linewidths in units of $\Delta\omega$ from the linewidths in units of the gate voltage the proportionality factor of the linear DC Stark shift is used. As in ref [43], a constant factor $c = 0.78\,GHz$ had to be added in order to fit the FWHM curve. This constant accounts for the charge fluctuations in the QD environment resulting in the fluctuation of the resonance energy. From the fits the obtained values are $\Gamma = 2.0\,GHz$ and $P_{sat} = 1.2\,nA$ in units of the measured photocurrent. This photocurrent corresponds to a laser power of $\sim 0.9\,nW$, assuming the room temperature quantum efficiency for the photo diode. The quantum efficiency at $4\,K$ decreases by couple of factors but we did not characterize it for the particular measurement.

5.3. Optical cooling of a single electron spin

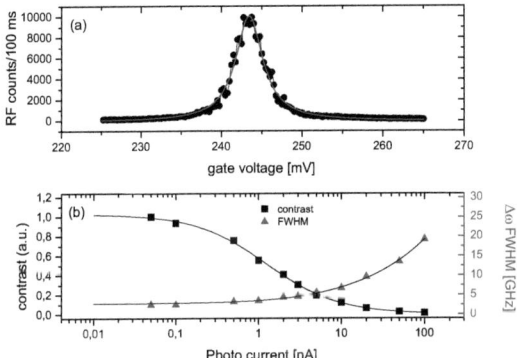

Figure 5.3.: Resonance fluorescence (RF) counts as a function of gate voltage at $B = 0\,T$ when the QD is excited by a laser with power $P_L^{det} \approx P_{sat}$. The curve is fitted to a Lorentzian (top). The contrast P_{scat}^{det}/P_L^{det} and FWHM of the linewidth $\Delta\omega$ as a function of detector photo current. The data is fitted to 5.34 and 5.33 given in the text (bottom).

5.3. Optical cooling of a single electron spin

The diagonal decay rate γ is responsible for the population transfer from the state $|1\rangle$ to the state $|2\rangle$, if the laser drives the transition $|1\rangle \leftrightarrow |3\rangle$. It has already been noted that the origin of the diagonal decay rate γ is the heavy hole-light hole mixing for fields larger than $200\,mT$ at which the hyperfine interaction induced ground state mixing can be neglected. Another contribution to γ might come from an imperfect alignment of the sample normal to the B–field, resulting in ground state mixing. Define z as the normal to the sample surface. In 5.1 $\mathbf{B} = B\hat{z}$ is implicitly assumed. For a given sample tilt the following terms in the Hamiltonian 5.1

$$\frac{\mu_B}{2}g_e\hat{\sigma}_{z,e}B + \frac{\mu_B}{2}g_h\hat{\sigma}_{z,h}B + \hbar\Omega_N(|1\rangle\langle 2| + |2\rangle\langle 1|)$$

are replaced by the following terms:

$$\frac{\mu_B}{2}\hat{\sigma}_e\mathbf{B}_e + \frac{\mu_B}{2}\hat{\sigma}_h\mathbf{B}_h \quad (5.35)$$

where $\hat{\sigma}_e$ ($\hat{\sigma}_h$) is the Pauli vector for the ground (trion) states. The effective magnetic fields $\mathbf{B}_e$ for the electron and $\mathbf{B}_h$ for the hole are defined as

$$\mathbf{B}_e = \begin{pmatrix} g_{ex}(\mathbf{B}_{ext} + \mathbf{B}_N^e)_x \\ g_{ey}(\mathbf{B}_{ext} + \mathbf{B}_N^e)_y \\ g_{ez}(\mathbf{B}_{ext} + \mathbf{B}_N^e)_z \end{pmatrix} \,;\, \mathbf{B}_h = \begin{pmatrix} g_{hx}(\mathbf{B}_{ext} + \mathbf{B}_N^h)_x \\ g_{hy}(\mathbf{B}_{ext} + \mathbf{B}_N^h)_y \\ g_{hz}(\mathbf{B}_{ext} + \mathbf{B}_N^h)_z \end{pmatrix} \quad (5.36)$$

where $\mathbf{B}_N^e$ ($\mathbf{B}_N^h$) is the effective nuclear field for the electron (hole) and the notation is such that g_{ex} is the electron g–factor in the x-direction. In general the effective fields $\mathbf{B}_e$ and $\mathbf{B}_h$ and therefore the electron and hole eigenstates point in different directions, both deviating from $\hat{z}$. Denoting the electron (trion) eigenstates in the presence of $\mathbf{B}_e$ ($\mathbf{B}_h$) as $|\tilde{1}\rangle, |\tilde{2}\rangle$ ($|\tilde{3}\rangle, |\tilde{4}\rangle$), the Liouvillian can be written in these basis and the spin pumping rate can be obtained from the following equation

$$\langle \tilde{2}|\rho|\tilde{2}\rangle = \langle \tilde{3}|\rho|\tilde{3}\rangle \, \Gamma \left\{ \left(\frac{\theta_h^2}{4} + \frac{\theta_e^2}{4}\right) + \eta_{hh-lh}\left(1 - \frac{\theta_h^2}{4} - \frac{\theta_e^2}{4}\right) \right\} \quad (5.37)$$

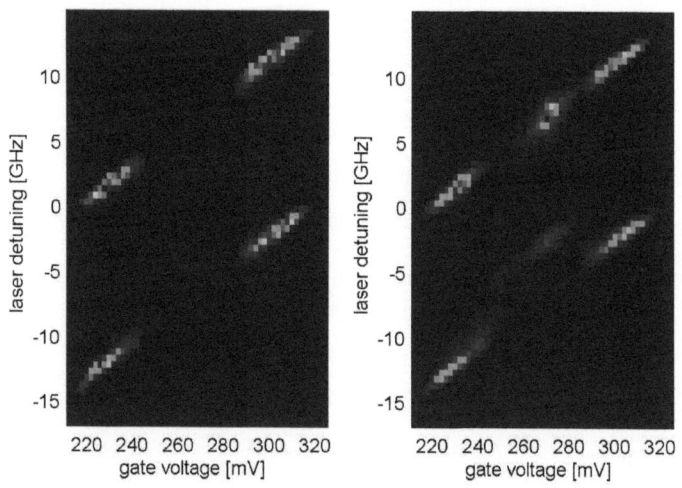

Figure 5.4.: (a) Resonance fluorescence data as a function of laser detuning $\frac{\Delta\omega}{2\pi}$ and gate voltage. At $B = 500\,mT$ a linearly polarized laser is scanned across the QD transitions and the gate voltage is stepped within the single electron regime. The signals from the transitions $|\uparrow\rangle \leftrightarrow |\downarrow\uparrow\Uparrow\rangle$ (above $\frac{\Delta\omega}{2\pi} = 0\,GHz$) and $|\downarrow\rangle \leftrightarrow |\downarrow\uparrow\Downarrow\rangle$ (below $\frac{\Delta\omega}{2\pi} = 0\,GHz$) are visible in some gate voltage ranges. Since, apart at the edges of the gate voltage plateaus, the cotunneling rate κ is small, the signals disappear due to electron spin pumping (left). A second linearly polarized laser is introduced which at a gate voltage of $270\,mV$ becomes on resonance with the $|\downarrow\rangle \leftrightarrow |\downarrow\uparrow\Downarrow\rangle$ transition. When the first laser hits the resonance $|\uparrow\rangle \leftrightarrow |\downarrow\uparrow\Uparrow\rangle$ at $\frac{\Delta\omega}{2\pi} \approx 7\,GHz$ ($|\uparrow\rangle \leftrightarrow |\downarrow\uparrow\Downarrow\rangle$ at $\frac{\Delta\omega}{2\pi} \approx -2.5\,GHz$) the double resonance condition leads to the recovery (partial recovery) of the signal.

Here θ_e (θ_h) is the angle between $\mathbf{B}_e$ ($\mathbf{B}_h$) and $\hat{z}$ and terms up to order θ^2 are kept. It is expected that $\frac{g_{hx}}{g_{hz}} < \frac{g_{ex}}{g_{ez}}$ and hence $\theta_h < \theta_e$. Also in-plane electron g−factor is expected to be slightly less negative than the vertical g-factor ($g_{ex} \sim -0.5$, $g_{ez} \sim -0.6$) [44–46]. Therefore an upper bound for the effective branching ratio can be written as

$$\eta_{eff} = \frac{\theta^2}{2} + \eta_{hh-lh} \quad (5.38)$$

where θ is the tilt angle of the sample. As will be shown in the next chapter, we measured η_{eff} to be $\frac{1}{250}$ for 10 different QDs. If it were determined by the sample

5.3. Optical cooling of a single electron spin   17

tilt only, $\theta = \sqrt{\frac{2}{125}}\frac{180^0}{\pi} \approx 5^0$. Our sample alignment with respect to the z−axis is expected to be better than 5^0. On the other hand, η_{hh-lh} is expected to deviate for different QDs [45, 46]. Since a distribution in η_{eff} is not observed the possibility that η_{eff} is determined by the sample tilt is not excluded.

When the magnetic field is large enough such that the coupling of the laser to the diagonal transitions can be neglected, as a result of the diagonal decay channel the laser driving the $|1\rangle \leftrightarrow |3\rangle$ transition prepares the electron in $|\downarrow\rangle$-state with a fidelity close to 100%, given that $\kappa << \eta \rho_{33}$ [11].

The left panel of fig.5.4 shows the RF data at $B = 500\,mT$ when a linearly polarized laser is scanned and the gate voltage is stepped within the single electron regime. The zero laser detuning is defined as the mean of the two vertical transitions in the middle of the plateau. For gate voltages at the edge of the single electron plateau the cotunneling rate κ is large and when the laser hits either the $|1\rangle \leftrightarrow |3\rangle$ or $|2\rangle \leftrightarrow |4\rangle$ transition an increase in the RF counts is observed. In the middle of the plateau κ is small and as a result of spin pumping the signals disappear. In the right panel of fig.5.4 a second linearly polarized laser is introduced which is kept at constant frequency, and which is on resonance with the $|2\rangle \leftrightarrow |4\rangle$ transition at a gate voltage of $270\,mV$. Therefore at this gate voltage a bidirectional spin pumping occurs when the first laser is on resonance with the $|1\rangle \leftrightarrow |3\rangle$ transition at around $7\,GHz$ detuning. Because of the bidirectionality a signal can be observed. Around $-3\,GHz$ detuning when the first laser hits the $|1\rangle \leftrightarrow |4\rangle$ resonance the observed signal is by about a factor of 6 smaller. This has two reasons. For the resonance condition of $|1\rangle \leftrightarrow |3\rangle$, RF photons are measured both from the $|1\rangle \leftrightarrow |3\rangle$ and $|2\rangle \leftrightarrow |4\rangle$ transitions. On the other hand for the resonance condition of $|1\rangle \leftrightarrow |4\rangle$ RF is only coming from the $|2\rangle \leftrightarrow |4\rangle$ transition. We would expect a factor of 2 difference only if the spin pumping rates into the $|\downarrow\rangle$ state were equal for both resonance conditions.

Due to the nature of the hh-lh mixing the spin pumping rates can differ for these two resonance conditions. An optically generated (pseudo)spin-down hh is mixed with lh states as $|\Downarrow\rangle_{hh} \approx |\Downarrow\rangle_{hh} + \epsilon_+|\Uparrow\rangle_{lh} + \epsilon_-|\Downarrow\rangle_{lh}$. (The corresponding mixing exists for the (pseudo)spin-up hh.) The spin-up(down) lh state has the angular momentum of $\frac{1}{2}(-\frac{1}{2})$ whereas the spin-up(down) hh state has the angular momentum of $\frac{3}{2}(-\frac{3}{2})$. Due to optical selection rules left-hand circularly polarized(σ^-) light couples to the diagonal transition via the ϵ_- mixing ($|\uparrow\rangle \rightarrow |\Downarrow\rangle_{lh}$) whereas the light coupling due to ϵ_+ mixing ($|\uparrow\rangle \rightarrow |\Uparrow\rangle_{lh}$) is polarized perpendicular to the sample plane and does not propagate in the z-direction. For the resonance condition of $|1\rangle \leftrightarrow |3\rangle$ the path for the spin pumping into the $|\downarrow\rangle$ state is $|1\rangle \xrightarrow{\frac{\Omega^2}{\Gamma}} |3\rangle \xrightarrow{(|\epsilon_+|^2+|\epsilon_-|^2)\Gamma} |2\rangle$. On the other hand for the diagonal resonance condition of $|1\rangle \leftrightarrow |4\rangle$ the path for the spin pumping into the $|\downarrow\rangle$ state is $|1\rangle \xrightarrow{\frac{(\Omega|\epsilon_-|)^2}{\Gamma}} |4\rangle \xrightarrow{\Gamma} |2\rangle$. Here Ω is the Rabi frequency due to the σ^- polarized component of the linearly polarized laser. The spin pumping rate in the case of the resonance condition $|1\rangle \leftrightarrow |3\rangle$ is larger by a factor of $\frac{\frac{\Omega^2}{\Gamma}\cdot(|\epsilon_+|^2+|\epsilon_-|^2)\Gamma}{\frac{(\Omega|\epsilon_-|)^2}{\Gamma}\cdot\Gamma} = \frac{|\epsilon_+|^2+|\epsilon_-|^2}{|\epsilon_-|^2}$.

Fig. 5.5 is a cross section of the left panel of fig.5.4 at $225\,mV$ gate voltage. Here the diagonal coupling $|1\rangle \leftrightarrow |4\rangle$ is directly observed. Note that the ratio of the peak

18   Resonantly probing a single electron spin

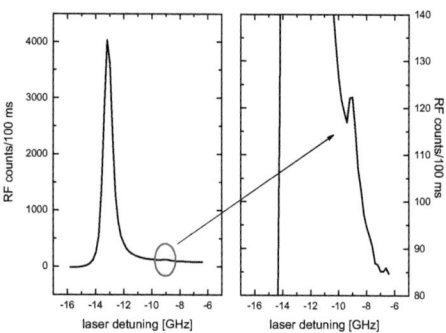

Figure 5.5.: Cross section of the left panel of 5.4 showing a laser scan at a gate voltage of $225\,mV$ (left) and it's zoom-in (right). The large peak is the resonance with $|\downarrow\rangle \leftrightarrow |\downarrow\uparrow\downarrow\rangle$ transition and the small signal at $\frac{\Delta\omega}{2\pi} = -9\,GHz$ is the resonance with the diagonal transition $|\uparrow\rangle \leftrightarrow |\downarrow\uparrow\downarrow\rangle$.

sizes is not necessarily η_{eff} since the photons coming from the diagonal decay might have an arbitrary polarization and the driving laser polarization couples only partly to this transition.

5.4. Signal-to-noise ratio analysis for time-resolved electron spin measurement

In a single shot spin measurement the aim is to obtain a spin dependent signal in a time interval before an environment or measurement induced spin flip occurs. If the signal-to-noise ratio (SNR) is larger than 1, spin quantum jumps can be observed. For smaller SNR, starting with the same initial conditions, the measurement can be repeated n-times until SNR of the measurement average exceeds 1 such that an ensemble averaged signal can be observed. In the case of the RF measurement the signal for the single-shot spin measurement is

$$S_{RF} = \dot{n}_{RF}\Delta t \tag{5.39}$$

and the photon shot noise given by

$$N_{RF} = \sqrt{\dot{n}_{RF}\Delta t} \tag{5.40}$$

where $\dot{n}_{RF}$ is the detected photon rate and Δt is the integration time.

$$\Longrightarrow SNR_{RF} = \sqrt{\dot{n}_{RF}\Delta t} \tag{5.41}$$

For a given collection efficiency $n_{RF} \propto \rho_{33}$. Setting $\Delta t = \tau_{spin}$ (spin life time) and given that the spin life time is limited by laser induced spin flips,

5.4. Signal-to-noise ratio analysis for time-resolved electron spin measurement

$$\tau_{spin} = \frac{1}{\eta \Gamma \rho_{33}} \quad (5.42)$$

$$\Rightarrow \Delta t \propto \frac{1}{n_{RF}} \quad (5.43)$$

This relation suggests that SNR is independent of laser excitation intensity. Therefore, the general expression for SNR_{RF} can be written in the low power limit. For the RF detection a linear polarizer has to be used which will suppress the QD-photons by a factor of 2, reducing the SNR_{RF} by a factor of $\sqrt{2}$. Assuming also in this limit $\Gamma_{lw} = \Gamma$ (negligible additional broadening) [43]

$$SNR_{RF} = \frac{1}{\sqrt{2}} \sqrt{\frac{2\epsilon_0 c}{\hbar \omega_L} \int_{det} \langle \hat{\mathbf{E}}_{QD}^- \hat{\mathbf{E}}_{QD}^+ \rangle ds} \sqrt{\Delta t} \quad (5.44)$$

$$\approx \frac{1}{\sqrt{2}} \sqrt{\frac{2\epsilon_0 c}{\hbar \omega_L} \int_{det} \langle \hat{\mathbf{E}}_{QD}^- \rangle \langle \hat{\mathbf{E}}_{QD}^+ \rangle ds} \sqrt{\Delta t} \quad (5.45)$$

When the absorption in an interference experiment is measured, the signal is (in units of photon number)

$$S_{abs} = \int_0^{\Delta t} \frac{P_{abs}}{\hbar \omega_L} dt = \Delta t \frac{2\epsilon_0 c}{\hbar \omega_L} \int_{det} (\mathbf{E}_L^- \mathbf{E}_{QD}^+ + \mathbf{E}_L^+ \mathbf{E}_{QD}^-) ds \quad (5.46)$$

The noise is the shot noise of the laser given by

$$N_{abs} = \sqrt{\Delta t \frac{2\epsilon_0 c}{\hbar \omega_L} \int_{det} (\mathbf{E}_L^- \mathbf{E}_L^+) ds} \quad (5.47)$$

The signal, for a given laser intensity, is maximum when for the ideal case the laser has the same spatial profile and polarization as the QD field, i.e. $\alpha \mathbf{E}_{QD} = \mathbf{E}_L$. Consider also the situation $\Delta \omega = 0$ i.e $Re\{\rho_{31}\} = 0$ and α to be reel. In this limit of maximum SNR_{abs} the expression reduces to

$$SNR_{abs} = \sqrt{\frac{2\epsilon_0 c}{\hbar \omega_L} \Delta t} \frac{\int 2\alpha \mathbf{E}_{QD}^- \mathbf{E}_{QD}^+ ds}{\sqrt{\int \alpha^2 \mathbf{E}_{QD}^- \mathbf{E}_{QD}^+ ds}}$$

$$= 2\sqrt{\frac{2\epsilon_0 c}{\hbar \omega_L} \Delta t \int \mathbf{E}_{QD}^- \mathbf{E}_{QD}^+ ds} \quad (5.48)$$

which is $2\sqrt{2}$ times the value of SNR_{RF}. The factor of 2 is coming from the factor of 2 in the interference, i.e. $S_{abs} \propto (\mathbf{E}_L^- \mathbf{E}_{QD}^+ + \mathbf{E}_L^+ \mathbf{E}_{QD}^-) = 2Re\{\mathbf{E}_L^- \mathbf{E}_{QD}^+\}$. The factor of $\sqrt{2}$ is originating from the fact that the collection polarizer absorbs half of the circularly polarized RF photons. Since there is no perfect overlap between the Gaussian laser beam and the QD field, a background-free time-resolved RF spin detection will in practice perform better than the calculated ratio, if the collection efficiencies are the same.

5.5. Measurement in other polarization basis

A general measurement method can be considered where the driving laser has an arbitrary polarization and the composite field $\mathbf{E}_L + \mathbf{E}_{QD}$ is detected in an arbitrary basis. An expression for such a measurement will be obtained and the RF and the absorption measurements will be shown to be special cases. Before proceeding a simplified expression for the signal $2\epsilon_0 c \int_{det}[(\mathbf{E}_L^- + \mathbf{E}_{QD}^-)(\mathbf{E}_L^+ + \mathbf{E}_{QD}^+)]\hat{\mathbf{u}}_c d\mathbf{s}$ will be given by making some approximations. First note that both the QD and the laser has phase profiles of a spherical wave in the far field so that the relative phase difference is independent of $\mathbf{r}$. Before doing any polarization operation on the composite field, a circularly symmetric objective collimates the beams such that afterward the z-component of the polarization can be considered to be zero. The dot field is σ^+ polarized and 5.1 shows that any mixture with the σ^- component will be at least a factor of $\frac{1}{\theta_{collect}^2}$ smaller, where $\theta_{collect}$ is the maximum collection angle given in chapter 1. This contribution (≈ 0.1) is neglected here. Therefore at the location of the QD ($x = 0, y = 0, z = 0$) and at the location of the polarization operation the polarizations are the same. With these considerations, as far as spatial variations are concerned, the vectorial nature of the fields can be reduced to constant polarization states in the lateral plane in the far field and for every point $\mathbf{r}$ in the far field, the QD field can be written as

$$E_{QD}^+(\mathbf{r}) = (iD(\mathbf{r}) - A(\mathbf{r}))\langle \sigma^+ | E_L^+(\mathbf{r}) \rangle |\sigma^+\rangle \quad (5.49)$$

where A and D are reel and proportional to the real and imaginary parts of ρ_{31}, i.e

$$D = \epsilon Re\{\rho_{31}\} \quad (5.50)$$
$$A = \epsilon Im\{\rho_{31}\} \quad (5.51)$$

$\epsilon < 1$ depends on the collection efficiency and the QD field-laser field overlap integral. Only the magnitudes of A and D change with $\mathbf{r}$. The laser field has the general polarization written as

$$E_L^+ = \sqrt{\frac{n}{2}} e^{i\omega_L t}(\cos \alpha e^{i\beta/2}|\sigma^+\rangle + \sin \alpha e^{-i\beta/2}|\sigma^-\rangle) \quad (5.52)$$

The dot field under this excitation is

$$E_{QD}^+ = \sqrt{\frac{n}{2}} e^{i\omega_L t}(\cos \alpha e^{i\beta/2}(-A + iD)|\sigma^+\rangle) \quad (5.53)$$

Here the convention is

$$dP_L = I_L ds = (2E_L^+ E_L^-)ds = n \quad (5.54)$$

with ds the unit area at the position $\mathbf{r}$ on the detector and n is the number of collected photons per unit time in this area. The detection is done in an arbitrary polarization basis $|\psi^\pm\rangle$ in the Poincaré sphere (fig. 5.6) given by

5.5. Measurement in other polarization basis

$$|\psi^+\rangle = \cos\frac{\theta}{2}|x\rangle + e^{i\phi}\sin\frac{\theta}{2}|y\rangle \quad (5.55)$$

$$|\psi^-\rangle = \sin\frac{\theta}{2}e^{-i\phi}|x\rangle + \cos\frac{\theta}{2}|y\rangle \quad (5.56)$$

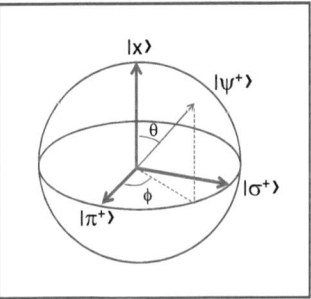

Figure 5.6.: Poincare sphere showing the arbitrary measurement basis state $|\psi^+\rangle$.

In order to calculate the signal $dS = 2|\langle\psi^+|E_L^+ + E_{QD}^+\rangle|^2 ds$, E_L and E_{QD} can be written in the basis of $|\psi^\pm\rangle$. If both the excitation and detection basis are $|\sigma^+\rangle$, then $\alpha = 0, \beta = 0, \theta = \frac{\pi}{2}, \phi = \frac{\pi}{2}$ and $2|\langle\psi^+|E_L^+ + E_{QD}^+\rangle|^2 = n(1 - 2A + A^2 + D^2)$, where the first term is the laser background, the second term the absorption signal and the last terms are for the RF photons. This the conventional absorption experiment with $A, D \ll 1$. In case a linearly polarized laser makes angle θ_L to $|x\rangle$ and the detection is in $|y\rangle$, then $\alpha = \frac{\pi}{4}, \beta = -2\theta_L, \theta = \pi, \phi = 0$ and

$$2|\langle\psi^+|E_L^+ + E_{QD}^+\rangle|^2 = \frac{n}{4}(2 + A^2 - 2A + 2A\cos(2\theta_L) - 2\cos(2\theta_L) + D^2 - 2D\sin(2\theta_L)) \quad (5.57)$$

For $\theta_L = 0$ (cross-polarized detection) $2|\langle\psi^+|E_L^+ + E_{QD}^+\rangle|^2 = \frac{n}{4}(A^2 + D^2)$ which is the usual RF signal. The factor $\frac{1}{4}$ is originates from the fact that only σ^+-component of the linearly polarized laser drives the QD and the collection polarizer cuts half of the σ^+-polarized QD photons. For $\theta_L \ll 1$ an interesting situation occurs. Keeping terms up to second order in θ_L, A and D

$$2|\langle\psi^+|E_L^+ + E_{QD}^+\rangle|^2 = \frac{n}{4}(A^2 + D^2 + 4\theta_L^2 - 4D\theta_L) \quad (5.58)$$

Therefore this technique allows the detection of $Re\{\rho_{31}\}$, i.e. the dispersive part of the QD response.

Such a dispersive signal is shown in fig. 5.7 with $\theta_L = 0.01$. APD counts are plotted as a function of the gate voltage. The magnetic field is $1T$ and the signal from $|\downarrow\rangle \leftrightarrow |\uparrow\downarrow\Downarrow\rangle$ transition is observed. Since the detected power is low, the measurement is shot noise limited without the need of lock-in modulation. On the other hand, shot noise limited absorption measurements (apart from extremely low excitation powers so that APDs can be used) rely on lock-in modulation, due to the

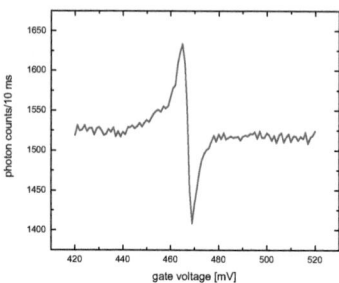

Figure 5.7.: (a) Gate voltage scan at $B = 1\,T$ showing the dispersive signal from the $|\downarrow\rangle \leftrightarrow |\downarrow\uparrow\Downarrow\rangle$ transition. The linearly polarized excitation laser makes an angle $\theta_L = 0.01$ to $|x\rangle$. The reflected laser photons together with the dot photons are sent to an APD after passing through a polarizer oriented along $|y\rangle$.

electrical noise of the photodiode at moderate powers and due to the classical laser noise at high powers. It can be shown that the SNR for single shot electron spin measurement with an ideal dispersive signal is half of that of an ideal absorption experiment. The insensitivity of the dispersive signal to fluctuations in detuning can be advantageous for spin measurement. Unlike the RF or the absorption measurement the sign of the dispersion signal is spin state dependent, i.e. if the signal is positive for $|\uparrow\rangle$ it will be negative for $|\downarrow\rangle$. Previously such a dispersive signal has been measured with the lock-in technique using two photodiodes and a polarizing beam cube [47].

Until now the signal at a given $\mathbf{r}$ has been calculated. In order to obtain an intuitive picture of the measurement the expression for the total detected signal will be simplified further. Assume a spherical detector positioned at a distance $|\mathbf{r}|$ from the QD. Accepting an error of the order of $\theta_{collect}^2 = 0.34^2 \approx 0.1$ allows to neglect the position dependence of $D(\mathbf{r})$ and $A(\mathbf{r})$ (see 5.1).

$$\Rightarrow \int_0^{\theta_{collect}} E_{QD}^+ E_L^- d\Omega \approx E_{QD}^+ \int_0^{\theta_{collect}} E_L^- d\Omega \tag{5.59}$$

Moreover

$$\frac{\int_0^{\theta_{collect}} E_L d\Omega}{\sqrt{\int_0^{\theta_{collect}} |E_L|^2 d\Omega}\sqrt{\int_0^{\theta_{collect}} d\Omega}} = 0.96 \approx 1 \tag{5.60}$$

allows to write

$$\int_0^{\theta_{collect}} E_L d\Omega \equiv E_{L,eff} \int_0^{\theta_{collect}} d\Omega \tag{5.61}$$

$$\int_0^{\theta_{collect}} |E_L|^2 d\Omega \approx |E_{L,eff}|^2 \int_0^{\theta_{collect}} d\Omega \tag{5.62}$$

5.5. Measurement in other polarization basis

In all the analysis for the signals and noises given in this chapter these relations allow to consider the fields to be constants over the detector surface given by $E_L = E_{L,eff}$ and $E_{QD}^\pm = E_{L,eff}^\pm(\pm iD - A)$ with D and A independent of $\mathbf{r}$.

6. Quantum-dot-spin single-photon interface

6.1. Introduction

As we saw in section 5.1 optical excitations of single electron charged quantum dots (QD) have favorable selection rules that allow for recycling trion transitions where the scattered photon polarization is strongly correlated with the electron spin-state [9]. However the realization of a spin-photon interface in the spirit of what has been recently realized for trapped ions [48] and recently for a single electronic spin of a nitrogen vacancy centre in diamond [49] suffers from the fact that the background excitation laser scattering from the solid-state interfaces and defects overwhelms the QD resonance fluorescence. While single QD resonance fluorescence (RF) has been recently observed by several groups [50, 51], the reported experiments did not demonstrate a complete suppression of the background in a charge-controlled QD that is essential for the realization of a spin-photon interface.

Combining our high level laser suppression by up to a factor of 10^7 with our high overall photon collection efficiency of 0.1 % we realized a classical spin-photon interface. We show that detection of a single photon, resonantly scattered on the charged-exciton (trion) resonance, projects the QD spin to a state where the spin is pointing along the external magnetic field ($|\uparrow\rangle$) with a conditional initialization fidelity of 99.2%. The bunching of resonantly scattered photons reveals information about electron spin dynamics allowing us to measure directly the cotunneling rate.

6.2. Background-free resonance fluorescence measurement

To determine the transition energies of the QD differential transmission (DT) measurements are carried out. Here we measure the interference of the transmitted laser and the QD photons with a photodiode below the sample in the cryostat. The photocurrent is converted to voltage with a preamplifier outside of the cryostat. In order to suppress the low frequency noise we make use of the lock-in technique. The signal is modulated by modulating the sample gate voltage and the output of the preamplifier is sent to the lock-in amplifier for demodulation. In the experiments presented here, we have only addressed the higher energy (blue) trion transition $|\uparrow\rangle \leftrightarrow |\uparrow\downarrow\Uparrow\rangle$ with a resonant laser field; provided that $B > 0.2T$, excitation of the $|\downarrow\rangle \leftrightarrow |\downarrow\uparrow\Downarrow\rangle$ transition can be safely neglected. Fig. 6.1(b)&(c) show the DT signal from the trion transition with a 100 ms integration time at $B = 1T$ and $B = 0T$

This chapter is based on [14].

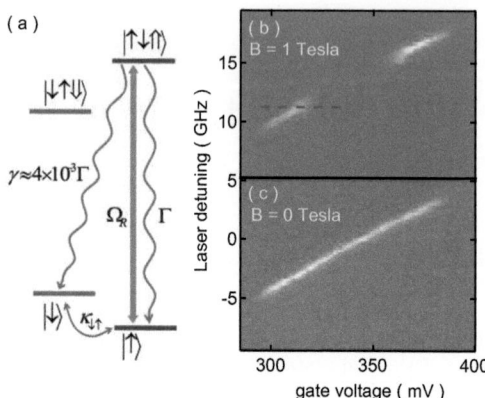

Figure 6.1.: (a) Energy level diagram for a quantum dot (QD) charged with a single electron. (b)&(c) Differential Transmission (DT) signal as a function of gate voltage and laser frequency at $B = 1T$ and $B = 0T$. At $B = 0T$ DT signal is seen at gate voltages where the QD is singly charged. At $B = 1T$ the DT signal (white points) in the middle of the plateau disappears due to spin pumping. Dashed line corresponds to the gate voltage trace in Fig. 6.2(a)

. In the middle of the single electron plateau the DT signal disappears at $B = 1T$ since in this gate voltage regime the rate of the laser induced spin pumping into $|\downarrow\rangle$ state far exceeds the co-tunneling rate $\kappa_{\uparrow\downarrow}$, resulting in high fidelity preparation of the electron in the $|\downarrow\rangle$ state within several microseconds [11]. At the plateau edges, high co-tunneling rate ensures the randomization of the electron spin population. The broadening of the absorption lineshape for gate voltages where the DT signal starts to disappear is due to dynamical nuclear spin polarization [17].

As described in section 4.2 RF is detected by collecting the emitted QD photons through the focusing objective (NA = 0.55) and a single mode fiber to an APD in the Geiger mode. A linear polarizer that is placed before the collection fiber and oriented orthogonal to the reflected laser polarization extinguishes the reflected laser photons by a factor exceeding 10^6 while eliminating only half of the circularly polarized RF photons. Fig. 6.2(a) shows the RF signal as the gate voltage is scanned along the dashed line in fig. 6.1(b) with an excitation laser power 10 times below the QD saturation. The deviation of the RF lineshape from the expected Lorentzian is most likely due to the nuclear spin polarization [17], since we observe that for $B = 0$, the RF lineshape has a perfect Lorentzian lineshape. At the gate voltage when the laser is on resonance with the $|\uparrow\rangle \leftrightarrow |\downarrow\uparrow\Uparrow\rangle$ transition, the ratio of the RF counts to the total photon counts is 0.996 ± 0.001. Taking into account the branching ratio, the fidelity of our conditional state initialization is 0.992 ± 0.001. Based on measured RF counts ($\sim 500,000$/sec) at $B = 0T$ for a laser power above QD saturation and using the measured trion lifetime of ~ 1nsec, we estimate an unprecedented overall QD RF collection efficiency of $\sim 0.1\%$.

6.2. Background-free resonance fluorescence measurement

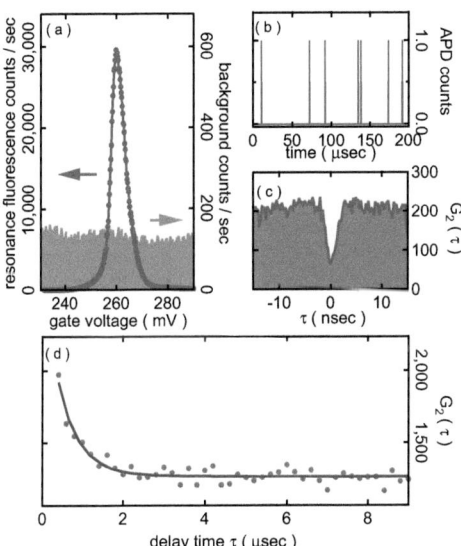

Figure 6.2.: (a) Resonance fluorescence (RF) signal from the QD as the gate voltage is scanned along the red dashed line on Fig. 6.1b. Laser power is well below the QD saturation, $P = 0.1 \cdot P_{sat}$. The gray trace shows the measurement background obtained at the same laser power with the laser frequency fully detuned from the QD transition. On resonance the ratio of RF photons to laser background exceeds 200. (b) A typical time trace recorded from the avalanche photo-diode (APD) with a 200 nsec time resolution with a resonant laser with $P = 0.1 \cdot P_{sat}$. Each pulse arises from the detection of a single photon, which indicates that the spin is in the $|\uparrow\rangle$ state with $(99.2 \pm 0.1)\%$ fidelity. (c) G_2 curve obtained by measuring photon coincidences on two APDs on nsec timescales. The expected antibunching behavior for a single emitter is observed, with a spontaneous emission rate $\Gamma \sim 10^9 s^{-1}$. The G_2 curve does not reach zero at $\tau = 0$ due to the finite time resolution $\sim 450\, psec$ of the Hanbury-Brown and Twiss measurement set-up. (d) Unnormalized photon correlation, G_2 curve obtained from $\sim 60,000$ traces such as the trace in (b) for $P = 0.1 \cdot P_{sat}$. Solid line is an exponential fit with a decay time $\tau_{decay} = (540 \pm 40)$nsec, corresponding to the cotunneling limited lifetime of the $|\uparrow\rangle$ state.

Fig. 6.2(b) shows a typical time trace of the APD counts integrated for 200 ns per point as the gate voltage in Fig. 6.2(a) is kept constant for resonant excitation. Whereas at a given time the absence of a photon detection provides almost no information about the electron spin-state, detection of an APD electrical pulse means with a confidence level of 99.2 % that the electron is in the $|\uparrow\rangle$ state. Moreover, we can locate the electron spin projection event onto the $|\uparrow\rangle$ state associated with each photon count to within 300psec of the arrival of the APD pulse; this projection time uncertainty is limited only by the jitter in the rise-time of the APD pulse.

6.3. Study of electron spin dynamics using photon correlation

We study RF count statistics by performing second order correlation (G_2) analysis. We first note that the RF photon correlations at short time-intervals exhibit the hallmark photon antibunching signature of a single quantum emitter (Fig. 6.2(c)). Fig. 6.2(d) shows the unnormalized photon correlation $G_2(\tau)$ curve obtained from $\sim 60,000$ time traces such as the one shown in Fig. 6.2(b), at the same gate voltage and laser power. Since laser photons follow Poissonian statistics with a flat $G_2(\tau)$ curve, the bunching behavior around $\tau = 0$ is a signature of RF photons, revealing information about the spin-flip dynamics. The decay of $G_2(\tau)$, fitted to an exponential function with the decay time $\tau_{decay} = (540 \pm 40)$ ns, provides a direct measurement of the $|\uparrow\rangle$ lifetime. As we discuss below, for the laser power used in this experiment, the laser induced spin decay rate is $(4.4\mu s)^{-1}$ and the spin lifetime is almost exclusively determined by co-tunneling processes. As will be derived in the next section, the cotunneling rate κ is half of the inverse of the $G_2(\tau)$-decay time. For the given gate voltage $\kappa = \frac{1}{2 \cdot 540} ns^{-1} = 9.3\, 10^5\, s^{-1}$. For the experimental conditions of fig. 6.2(b), we estimate that the conditional probability that we detect a photon given that the electron is in $|\uparrow\rangle$ state is $\sim (6.4 \pm 0.6)\%$. By varying the laser intensity the efficiency of spin-state detection can be improved to about 20%; this enhancement comes at the expense of an increase in the laser background such that only 98.6% of the detected photons originate from the QD RF.

In order to prove that the $G_2(\tau)$ decay time is indeed a measure of the spin lifetime, we perform an n-shot measurement [1] of the laser induced spin decay time in the middle of the single electron plateau where the co-tunneling rate is negligible and compare the result to the $G_2(\tau)$ measurements. The scheme of an n-shot measurement cycle is depicted in the inset of Fig. 6.3(a). In these n-shot spin measurements, a linearly polarized laser on resonance with the $|\uparrow\rangle \leftrightarrow |\uparrow\downarrow\Uparrow\rangle$ transition is switched on and over a time window of micro seconds the RF counts are saved with 200 ns integration time per data point. To undo spin-pumping that arises from the spin-measurement back-action in the form of spontaneous spin-flip Raman scattering into the $|\downarrow\rangle$ state, we apply a gate voltage pulse that first ejects the electron from the QD and then injects a new electron with a random spin. Repeating this cycle $\sim 10,000$ times at a laser power of $P_{sat} = 3nW$ corresponding to QD saturation, we obtain the exponentially decaying n-shot curve depicted in Fig. 6.3(a) (black points) with the decay time of $(860 \pm 20)ns$. The decay time of $G_2(\tau)$ obtained with the same laser power and at the same gate voltage is $(840\pm40)ns$ and is in good agreement with the n-shot measurement at the corresponding laser power (Fig. 6.3(a) (red points)). The laser induced decay rates are expected to be proportional to the trion population, which in turn is proportional to the RF counts. The measured decay times of $(480 \pm 30)ns$, $(860 \pm 20)ns$ and $(4.4 \pm 0.2)\mu s$ from the n-shot measurements at laser powers $10 P_{sat}$, P_{sat} and $0.1 \cdot P_{sat}$ agree well with the expected ratio of $1 : 2 : 10$ (data not shown). Fig. 6.3 (b),(c) show $G_2(\tau)$ measurements from another QD performed at two different gate voltages at the edge of the single electron plateau. In this regime the spin life time is determined by the cotunneling processes and the decay of the curves reveals the dependence of the cotunneling rate on the gate voltage.

6.4. Calculation of the 2nd order photon correlation $G_2(\tau)$

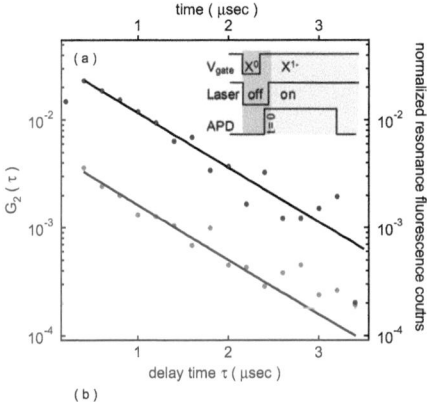

Figure 6.3.: (a) G_2 (red) and nshot (black) measurements obtained at a gate voltage in the middle of the plateau and laser intensity $P = P_{sat}$. Solid lines are exponential fits with decay times $(840 \pm 40)ns$ and $(860 \pm 20)ns$ respectively, demonstrating the agreement between the two measurement methods. (b)&(c) G_2 curves obtained from a second quantum dot with a laser power $P = 0.1 \cdot P_{sat}$ and at two gate voltages close to the plateau edge (large cotunneling) and $7.5mV$ apart. The solid lines are exponential fits, showing decay times of $(1.0\pm0.1)\mu s$ and $(2.5\pm0.2)\mu s$ respectively. These decay times are direct measurements of cotunneling-limited spin decay rates at the given gate voltages, with the faster decay, (b), corresponding to the gate voltage closer to the plateau edge as expected.

6.4. Calculation of the 2nd order photon correlation $G_2(\tau)$

In order to calculate $G_2(\tau)$ we will use the rate equations:

$$\dot{\rho}_{33} = R\rho_{11} - \Gamma(1+\eta)\rho_{33} \tag{6.1}$$

$$\dot{\rho}_{22} = \eta\Gamma\rho_{33} + \kappa\rho_{11} - \kappa\rho_{22} \tag{6.2}$$

$$\dot{\rho}_{11} = \Gamma\rho_{33} + \kappa\rho_{22} - (R+\kappa)\rho_{11} \tag{6.3}$$

with the condition

$$\rho_{11} + \rho_{22} + \rho_{33} = 1 \tag{6.4}$$

The levels are labeled as in chapter 5. Here κ is the bidirectional cotunneling rate, $\eta = \frac{\gamma}{\Gamma}$ is the branching ratio and R is the laser pumping rate. This equations can be written in the matrix form:

$$\frac{d}{dt}\begin{pmatrix}\rho_{11}\\\rho_{22}\\\rho_{33}\end{pmatrix} = M\begin{pmatrix}\rho_{11}\\\rho_{22}\\\rho_{33}\end{pmatrix} \tag{6.5}$$

$$M = \begin{bmatrix} -(R+\kappa) & \kappa & \Gamma \\ \kappa & -\kappa & \eta\Gamma \\ \kappa & 0 & -\Gamma(\eta+1) \end{bmatrix}$$

We first find the steady state solutions by setting $\frac{d}{dt}\begin{pmatrix}\rho_{11}\\\rho_{22}\\\rho_{33}\end{pmatrix} = 0$. The steady state solutions are:

$$\rho_{11} = \frac{1}{D}\kappa\Gamma(1+\eta) \tag{6.6}$$

$$\rho_{22} = \frac{1}{D}(R\eta\Gamma + \kappa\Gamma(1+\eta)) \tag{6.7}$$

$$\rho_{33} = \frac{1}{D}R\kappa \tag{6.8}$$

with D being the determinant of the matrix M:

$$D = R(\kappa + \eta\Gamma) + 2\kappa\Gamma(1+\eta). \tag{6.9}$$

For $\tau > 0$ the expression for $g_2(\tau)$ is given by

$$g_2(\tau) = \frac{\langle\hat{\sigma}_{31}(t)\hat{\sigma}_{31}(t+\tau)\hat{\sigma}_{13}(t+\tau)\hat{\sigma}_{13}(t)\rangle}{\langle\hat{\sigma}_{33}(t)\rangle^2} \tag{6.10}$$

$$= \frac{\langle\hat{\sigma}_{31}(t)\hat{\sigma}_{33}(t+\tau)\hat{\sigma}_{13}(t)\rangle}{\langle\hat{\sigma}_{33}(t)\rangle^2} \tag{6.11}$$

where $\hat{\sigma}_{ij} = |i\rangle\langle j|$ and the operators are in the Heisenberg picture. Using the quantum regression theorem [40] and 6.5 we can write the differential equation

$$\frac{d}{d\tau}\begin{pmatrix}\langle\hat{\sigma}_{31}(t)\hat{\sigma}_{11}(t+\tau)\hat{\sigma}_{13}(t)\rangle\\\langle\hat{\sigma}_{31}(t)\hat{\sigma}_{22}(t+\tau)\hat{\sigma}_{13}(t)\rangle\\\langle\hat{\sigma}_{31}(t)\hat{\sigma}_{33}(t+\tau)\hat{\sigma}_{13}(t)\rangle\end{pmatrix} = M\begin{pmatrix}\langle\hat{\sigma}_{31}(t)\hat{\sigma}_{11}(t+\tau)\hat{\sigma}_{13}(t)\rangle\\\langle\hat{\sigma}_{31}(t)\hat{\sigma}_{22}(t+\tau)\hat{\sigma}_{13}(t)\rangle\\\langle\hat{\sigma}_{31}(t)\hat{\sigma}_{33}(t+\tau)\hat{\sigma}_{13}(t)\rangle\end{pmatrix} \tag{6.12}$$

We let $t \to \infty$ and we write in short hand notation

$$\frac{d}{d\tau}\chi(\tau) = M\chi(\tau) \tag{6.13}$$

with the initial condition

$$\chi(0) = \begin{pmatrix}\rho_{33}\\0\\0\end{pmatrix} = \begin{pmatrix}\frac{R\kappa}{D}\\0\\0\end{pmatrix} \tag{6.14}$$

Comparing with 6.11 we see that the third column of χ will enable us to calculate $g_2(\tau)$. $\chi(\tau)$ can be calculated with the Laplace tranform

$$(s\mathbb{I} - M)\chi(s) = \chi(0) \tag{6.15}$$

After calculating $\chi(s)$, $g_2(\tau)$ is found to be

6.4. Calculation of the 2nd order photon correlation $G_2(\tau)$

$$g_2(\tau) = \frac{D}{\kappa}\left[\frac{\kappa}{s_1 s_2} + \frac{s_1+\kappa}{s_1(s_1-s_2)}e^{s_1\tau} + \frac{s_2+\kappa}{s_2(s_2-s_1)}e^{s_2\tau}\right] \quad (6.16)$$

with

$$s_{\frac{1}{2}} = -\frac{\Gamma(\eta+1)+R+2\kappa}{2} \mp \frac{1}{2}\sqrt{(\Gamma(\eta+1)-2\kappa-R)^2+4R(\Gamma-\kappa)} \quad (6.17)$$

The first two terms of 6.16 result in anti-bunching dip with a width of Γ, since for $R, \kappa \ll \Gamma$ $s_1 \approx -\Gamma(\eta+1)$. The third term is related to $|\uparrow\rangle$ life time. If the life time is limited by the cotunneling rate ($R \ll \kappa$), then we find $s_2 \approx -2\kappa$. This is the condition for Fig.6.2(d). If the measurement is done in the middle of X^1 plateau where the cotunneling rate κ is negligible ($\kappa \ll R$) then we find $s_2 \approx -R\eta$. Starting form the $|\uparrow\rangle$ state, on the time scale of Γ, the dot behaves like a two-level system (levels $|1\rangle$ and $|3\rangle$). Initially the excited state reaches the equilibrium value $\rho_{33} = \frac{R}{R+\Gamma} \approx \frac{R}{\Gamma}$ and the initial laser-induced spin flip rate is $\rho_{33}\eta\Gamma = R\eta$. This is also the decay rate of $g_2(\tau)$ when the cotunneling rate is neglible.

7. Measurement of heavy-hole hyperfine interaction in InGaAs quantum dots using resonance fluorescence

We measure the strength and the sign of hyperfine interaction of a heavy-hole with nuclear spins in single self-assembled quantum dots. Our experiments utilize the locking of a quantum dot resonance to an incident laser frequency to generate nuclear spin polarization. By monitoring the resulting Overhauser shift of optical transitions that are split either by electron or exciton Zeeman energy with respect to the locked transition using resonance fluorescence, we find that the ratio of the heavy-hole and electron hyperfine interactions is -0.09 ± 0.02 in three QDs. Since hyperfine interactions constitute the principal decoherence source for spin qubits, we expect our results to be important for efforts aimed at using heavy-hole spins in quantum information processing.

7.1. Introduction

Theoretical and experimental studies in the last decade have established that hyperfine interaction with the quantum dot (QD) nuclei constitute the principal decoherence mechanism for electron spin qubits [5, 6, 10, 52–54]. An interesting strategy to circumvent leakage of quantum information to the nuclear spin environment is to represent quantum information with the pseudo-spin of a QD heavy-hole (HH) [2, 3, 7]: since HH states are formed mainly out of bonding p-orbitals of the lattice atoms, it had been argued that the HH hyperfine interaction is negligible. Recently, it was shown theoretically that the hyperfine interaction of a HH with the nuclear spins in a QD, while being Ising-like, could in fact be comparable in strength to that of the electron [55, 56]. While prior experiments have addressed hole-nuclei coupling in an ensemble of QDs [57, 58], high resolution measurements of HH hyperfine interaction in single optically active QDs have remained unexplored.

In this chapter, we present resonance fluorescence (RF) measurements that directly reveal the relative strength of the HH-hyperfine interaction in single-electron charged InGaAs QDs. To this end, we generate a precise amount of nuclear spin polarization in an external magnetic field by dragging the higher energy (blue) Zeeman shifted charged exciton (trion) transition using a non-perturbative laser field [17]. We then measure the resulting Overhauser shift of the QD transitions that are shifted either by the Zeeman energy of the exciton (the red trion transition)

This chapter is based on [18].

or the electron (the forbidden/diagonal transition) with respect to the blue trion resonance. The nuclear spin polarization induced energy shifts in these transitions are determined by the difference and the sum of the Overhauser field seen by the electron and the HH, respectively. Measuring the nuclear spin polarization induced shift of these two transitions allows us to directly determine the ratio of the HH to electron Overhauser shift to be $\eta = -0.09 \pm 0.02$ in two QDs and $\eta = -0.10 \pm 0.05$ in a third QD.

7.2. Measurements

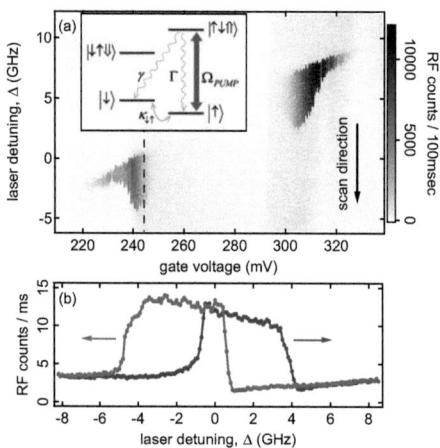

Figure 7.1.: (a) Resonance fluorescence (RF) signal from the blue trion transition vs gate voltage and pump laser detuning, Δ at $B = 4T$ and $P = P_{sat}/2$. Remainder of the experiments are performed at the gate voltage indicated by the dashed line, where the signal is reduced ~ 4 times due to spin pumping and a large line broadening due to dynamic nuclear spin polarization is observed. Inset: energy level diagram for a quantum dot charged with a single electron. (b) Cross section of (a) across the dashed line, opposite scan directions indicated by the arrows. A total dragging range of $\sim 8 GHz$ is observed. Interference with the laser background is partly responsible for the change of RF counts along the dragging range.

Our experiments combine two recent advances in experimental QD spin physics; namely the high-efficiency detection of RF [14, 50, 51] and the possibility to lock a QD resonance to an incident laser frequency via nuclear spin polarization [17]. Figure 7.1(a) shows the gate-voltage dependent RF signal from the blue trion transition at $B = 4T$ as the pump laser is scanned from an initial blue-detuning $(\omega_p > \omega_{t-b}^0)$ to a final red detuning $(\omega_p < \omega_{t-b}^0)$; here ω_{t-b}^0 is the trion resonance frequency in the absence of nuclear spin polarization and ω_p is the pump laser frequency. Reflected photons from the linearly polarized excitation laser are blocked by polarization sup-

7.2. Measurements

pression [14]. The RF signal is strong at the edges of the plateau where co-tunneling rate $\kappa_{\uparrow\downarrow}$ is large, and disappears in the middle of the plateau due to spin pumping [11, 16]. The characteristic extension of the peaks in the direction of the laser scan is due to dynamic nuclear spin polarization, i.e. the *resonance dragging* effect [17]. Experiments are performed at the gate voltage indicated by the dashed line where the dragging range is close to maximum. At this gate voltage the RF contrast is reduced ~ 4 times due to spin pumping. Figure7.1(b) shows laser scans obtained at a gate voltage denoted by the dashed line in Fig. 7.1(a) for two opposite scan directions: a total dragging range of ~ 8GHz is observed. We define the middle point between the onset of forward and backward dragging as $\omega_p - \omega_{t-b}^0 = \Delta = 0$. All other frequencies are measured relative to this point. In the remainder of the measurements we use the detuning Δ of the pump laser that is slowly scanned across the blue trion transition, in either forward or backward direction, as a knob for adjusting the amount of nuclear spin polarization in the QD.

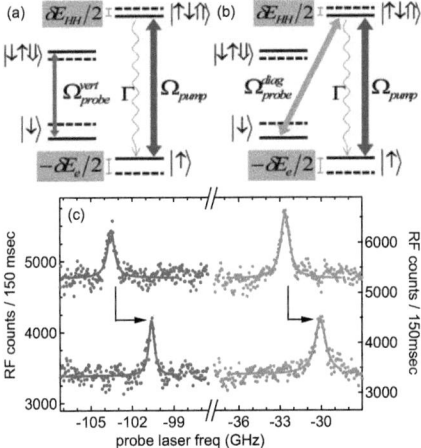

Figure 7.2.: (a)&(b) Energy level diagrams showing the pump and probe lasers. The probe laser re-pumps the spin into the $|\uparrow\rangle$ state by driving the red trion (a) or the diagonal (b) transition. Dashed (solid) lines indicate the ground and excited energy levels before (after) the polarization of nuclei by the pump laser. δE_{HH} (δE_e) denotes the Overhauser shift seen by a single QD HH (electron). (c) Resonance fluorescence (RF) signal recorded as the probe laser is scanned across the diagonal (right) or the red trion (left) transitions. An arbitrary offset is added to the top two scans for clarity. Prior to the probe laser scan the pump laser is scanned across the blue trion transition and stopped at a detuning of $\Delta = 0.5GHz$ (top) or $\Delta = -2.5GHz$ (bottom). Solid lines are Lorentzian fits. Peak positions are shifted due to nuclear spin polarization induced by the pump laser.

Since the electrons in the relevant optically excited states form a singlet, the nuclear-spin-polarization-induced change in the trion Zeeman splitting is given exclusively by the Overhauser shift of a single HH. To measure this HH Overhauser shift, we tune a strong pump laser to create a precise amount of nuclear spin polarization by dragging the blue trion transition from ω_{t-b}^0 to $\omega_{t-b}^0 + \Delta$. The pump laser which remains on resonance throughout the dragging range, at the same time leads to a substantial electron spin pumping into the $|\downarrow\rangle$ state, causing a reduction in the strength of the RF signal. Subsequently, we scan a weak probe laser across the red trion and the diagonal resonances: once the probe laser is on resonance with either the red trion (Fig. 7.2(a)) or diagonal (Fig 7.2(b)) [59] transition, it pumps the electron spin back to the $|\uparrow\rangle$ state, leading to a partial recovery of the RF signal. The top traces in Fig. 7.2(c) show the enhancement of the RF signal when the probe laser is on resonance with the red trion (left) or the diagonal (right) transitions, when the pump laser frequency was fixed at $\Delta = 0.5 GHz$. When we scan the pump laser to a final detuning of $\Delta = -2.5 GHz$, the resulting nuclear spin polarization modifies both the red trion and the diagonal transition resonance frequencies; the change in the red trion (diagonal) resonance is given by $(-\delta E_e(\Delta) + \delta E_{HH}(\Delta))/2$ $((-\delta E_e(\Delta) - \delta E_{HH}(\Delta))/2)$, where $\delta E_{HH}(\Delta)$ ($\delta E_e(\Delta)$) denotes the Δ-dependent Overhauser shift seen by a single QD HH (electron), i.e. the change in the Zeeman splitting of the HH (electron) in presence of nuclear spin polarization. Therefore, by measuring the shift in the corresponding resonances using the probe laser (Fig. 7.2(c) bottom trace), we determine the ratio $\eta(\Delta) = \delta E_{HH}(\Delta)/\delta E_e(\Delta)$ of the HH and electron Overhauser shifts to be -0.09. This reveals that the HH and the electron hyperfine interactions have opposite signs.

The shift in the red trion and the diagonal transition frequencies are extracted by fitting a Lorentzian lineshape to the resonantly enhanced RF signal (Fig. 7.2(c)). To confirm that the ratio $\eta(\Delta)$ is in fact independent of the amount of nuclear spin polarization, we repeat the experiment for a range of different Δ values. Figure 7.3(a) shows a series of probe scans across the diagonal transition for different pump laser detunings. The extracted transition energies are plotted in Figure 7.3(b) (green dots). The red dots in Fig. 7.3(b) show the shift of the red trion transition measured using the same technique. Dashed lines are linear fits to the extracted nuclear-spin-polarization-modified resonance frequencies.

The remarkable linear fit to the data in Fig. 7.3(b) not only shows that $\eta(\Delta) = \eta$ as we anticipated, but also that both δE_{HH} and δE_e scale linearly with the pump-laser detuning. This linear dependence allows us to simply determine the relative strength of the HH hyperfine interaction as $\eta = -(\alpha - \beta)/(\alpha + \beta)$, where α (β) is the slope of the red-trion (diagonal) transition depicted in Fig. 7.3(b). Using the measured slope of -0.98 ± 0.02 for the red line and -0.82 ± 0.02 for the green line we calculate $\eta = -0.09 \pm 0.02$.

An accurate determination of η requires that scanning the probe laser across the red trion or the diagonal transition does not modify the nuclear spin polarization that was created by the pump laser. Such a modification would occur if the probe laser itself caused dragging. If this were the case, the pump laser would no longer be resonant with the blue trion transition, leading to a decrease in the (pump laser induced) RF signal that constitutes the background of the scans shown in figure 7.2(c) [1]. We choose the probe laser power for both the diagonal and the red trion transi-

---

[1] Background is given by the pump laser induced RF signal when the probe laser is off resonance with the diagonal or the red trion transition.

7.2. Measurements

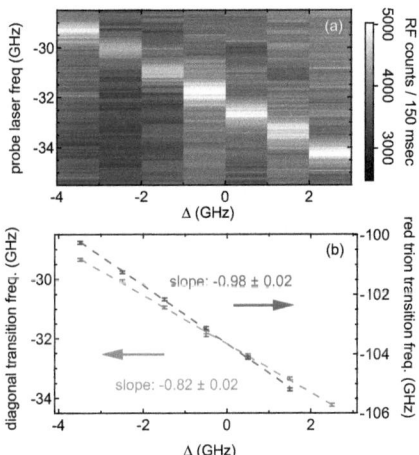

Figure 7.3.: (a) Resonance fluorescence (RF) signal as a function of probe laser frequency scanned across the diagonal transition and pump laser detuning, Δ. The position of the peak corresponding to recovered RF counts shifts due to nuclear spin polarization induced by the pump laser. (b) (green points) The position of the peaks in (a) extracted from Lorentzian fits to the data. Red points correspond to peak positions extracted from similar scans where the probe laser is scanned across the vertical transition. Dashed lines are linear fits with slopes indicated on the figure. Arrows point to the relevant axis.

tion such that there is no measurable change in the background as the probe laser is scanned across that transition. In addition, the linewidths of the recovered RF signal match the QD transition linewidth at $B = 0T$, in agreement with the absence of probe induced dragging. Furthermore we repeated the measurement at various probe laser powers; only when the laser power is low enough does the slope become completely independent of laser power. Figure 7.4(a) shows the Overhauser shifts of the vertical and diagonal transitions, each for two different probe laser powers indicated on the graph. The measured slopes at different powers are indeed in excellent agreement. The oscillator strength of the diagonal transition is much weaker than that of the red trion transition, requiring higher probe laser powers to induce resonant spin pumping. By repeating the diagonal transition Overhauser shift measurements for two different gate voltages, $1mV$ apart, around the gate voltage indicated by the dashed line on Fig. 7.1(b) (blue traces on Fig. 7.4(a)), we have also confirmed that the slopes do not depend on the gate voltage. The latter measurements also suggest that small changes in the charging environment and the resulting changes in the confined electron and HH wave-functions do not alter the ratio η. The ratio η also shows no appreciable dependence on the strength of the external magnetic field (Fig. 7.4(a) inset).

Due to the large variation in HH g-factor and the positively-charged trion con-

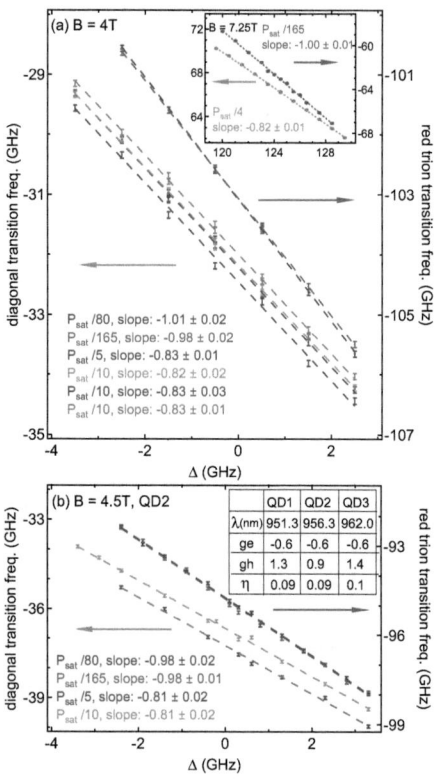

Figure 7.4.: (a) Transition energies of the red trion (red points) and the diagonal transition (green and blue points) vs pump laser detuning, Δ, at $B = 4T$. Probe laser power and slopes of linear fits (dashed lines) are indicated on the figure. The dark and light blue points are measured with the same laser power at two different gate voltages, $1mV$ apart, around the gate voltage indicated by the dashed line on Fig. 7.1(a). The dark and light green and red points are measured at two different laser powers at the gate voltage indicated by the dashed line on Fig. 7.1(a). The slopes do not change with changing probe laser power or gate voltage. Inset shows measurements at $B = 7.25T$: the axes are the same as those on the main figure. (b) Transition energies of the red trion (red points) and the diagonal transition (green points) vs Δ, measured on a second quantum dot. The vertical shift between the two green lines is due to a shift in the gate voltage, caused by fluctuations in the quantum dot charge environment. The box compares the emission wavelength, electron (ge) and heavy-hole (gh) g-factors and η for 3 different quantum dots.

finement energy, it is generally believed that the confined HH wave-function could change substantially from one QD to another. To determine if these changes lead

7.3. Nuclear spin polarization by driving the diagonal transition

to a modification of HH-hyperfine interaction, we have repeated our experiments on two other QDs. Figure 7.4(b) shows measurements on a second QD which yields $\eta = -0.09 \pm 0.02$. The ratio for a third QD, determined with a factor 2.5 lower accuracy, yielded $\eta \sim -0.1$. Remarkably, we find that the strength of the HH-hyperfine interaction in these 3 QDs to be almost identical, even though their HH longitudinal g-factors vary substantially (Fig. 7.4(b)).

7.3. Nuclear spin polarization by driving the diagonal transition

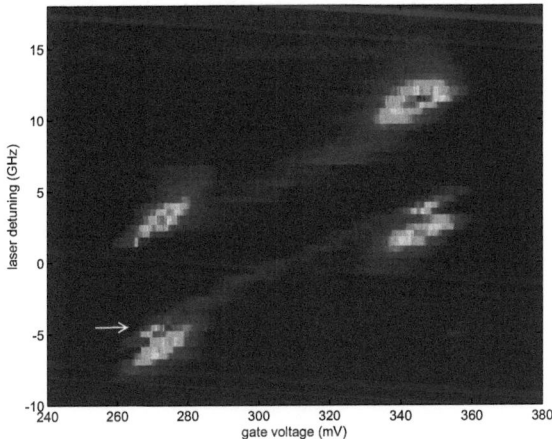

Figure 7.5.: Resonance fluorescence signal when the gate voltage is scanned in the positive direction for different laser frequencies. The magnetic field is $300\,mT$ and the excitation laser power $P \approx 10 P_{sat}$.

In this section we show an alternative method to create nuclear spin polarization by driving the weak diagonal transition. Fig. 7.5 shows resonance fluorescence signal when the gate voltage is scanned in the positive direction and the laser frequency is stepped. The magnetic field is $300\,mT$ and the excitation laser power $P \approx 10 P_{sat}$. This data is significantly different from Fig. 5.4 (a) where $P \approx 0.1 P_{sat}$ and the magnetic field is $500\,mT$. The additional signal indicated by the white arrow appears when the laser is on resonance with the diagonal transition $|\uparrow\rangle \leftrightarrow |\uparrow\downarrow\Uparrow\rangle$. Since this transition should be very weak, the signal is much larger than expected. On the other hand there is no signal when the laser is on resonance with the other diagonal transition $|\downarrow\rangle \leftrightarrow |\uparrow\downarrow\Downarrow\rangle$. We argue that when the laser drives the transition $|\uparrow\rangle \leftrightarrow |\uparrow\downarrow\Downarrow\rangle$ it creates nuclear spin polarization which almost cancels the electron Zeeman splitting. The laser then becomes quasi-resonant with the strong vertical transition $|\downarrow\rangle \leftrightarrow |\uparrow\downarrow\Downarrow\rangle$ resulting in a strong RF signal. On the other hand when the laser is driving the transition $|\downarrow\rangle \leftrightarrow |\uparrow\downarrow\Uparrow\rangle$ the nuclei are polarized in the opposite direction resulting in an increase of the electron Zeeman splitting.

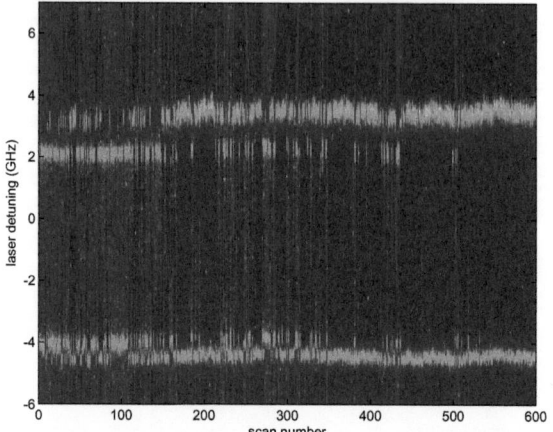

Figure 7.6.: Shown is differential reflection signal at $300\,mT$ from successive scans in the negative direction by the weak probe laser for a second QD. In order to create nuclear spin polarization, the gate voltage and the frequency of the strong linearly polarized pump laser is held at the position corresponding to the signal indicated by the arrow in Fig. 7.5. The pump laser is blocked in the collection path.

We test this hypothesis by performing a two-color differential reflection measurement on a second dot. The gate voltage and the frequency of the strong linearly polarized pump laser is held at a position corresponding to the signal indicated by the arrow in Fig. 7.5. The appearance of this signal means that the nuclei are polarized. We scan the frequency of the weak probe laser across the quantum dot transitions in the negative direction. The polarization of the probe laser is orthogonal to the pump laser and parallel to the collection polarizer (Fig. 4.4). Similar to the differential transmission (DT) measurement described in section 6.2 we do differential reflection (DR) measurements using gate modulation were the signal is the interference of the QD photons with the reflected instead of the transmitted laser light. Fig. 7.6 shows the DR-signal for successive probe laser scans. The lower (upper) peaks are from the red (blue) vertical transition. A bistable behavior is observed for both transitions revealing bistability of the nuclear spin polarization. When the nuclei are polarized the red (blue) transition is shifted toward blue (red). In some scans the pump laser cannot polarize the nuclei and sometimes the spin polarization is lost during the scan. The hyperfine induced shift of the red vertical transitions seems to be smaller than the shift of the blue transition. This is due to the AC Stark shift since the strong pump laser is quasi-resonant with the transition $|\downarrow\rangle \leftrightarrow |\uparrow\downarrow\Downarrow\rangle$ when there is nuclear spin polarization. In this scheme we could not figure out how much nuclear spin polarization is created and thus using this technique for hole hyperfine measurement was not feasible.

7.4. Conclusions

Despite the accurate measurement of the Overhauser shift of the HH and the electron that we have demonstrated, it is not straightforward to use our data to extract the actual HH hyperfine interaction constant with high accuracy due to differences in the confined electron and the HH envelope wave-functions. The exact mechanism behind dragging of QD resonances is not well understood; however, it is safe to assume that the underlying nuclear spin polarization is mainly mediated by the electron [2]. The precise magnitude of the HH Overhauser shift is therefore influenced by the overlap between the electron and hole wave-functions and their confinement length-scales. Repeating the experiments on different QDs, with vastly different in-plane g-factors, would yield further information about the sensitivity of η to the IIII confinement.

The striking feature of our experiments is the (almost) perfect linear dependence of the Overhauser shift on the pump-laser detuning. What is more remarkable is the fact that the slope of the red trion energy shift is ~ -1, indicating that the Overhauser shift of the blue trion transition satisfies $(-\delta E_e(\Delta)+\delta E_{HH}(\Delta))/2 = -\Delta+c$, where c is a constant much smaller than the bare optical transition linewidth. On the other hand theoretical models proposed to explain dragging showed a finite dependence of the absorption contrast on the bare laser detuning [17, 60], suggesting that the amount of nuclear spin polarization has a non-trivial dependence on the laser frequency. Our experiments could help in identifying the mechanisms underlying the dragging of QD resonances.

In summary, we have developed a new measurement technique combining two recent advances in QD physics to determine the strength of HH hyperfine interaction. Our measurements on highly strained self-assembled QDs indicate a coupling strength that is ~ 2 smaller than the theoretical prediction for strain-free GaAs QDs [55], and provide further support for efforts aimed at using confined HH pseudo-spins as qubits in solid-state quantum information processing. After completion of this work, we became aware of related experiments by Chekhovic et al. [62] on InP QDs.

---

[2]We note that W.Yang et al. [60] and X. Xu et al. [61] propose a dragging mechanism based on the HH hyperfine interaction.

8. Outlook

This thesis presents resonance fluorescence as a powerful tool for QD spectroscopy with large bandwidth and low noise. Some important future QD experiments which were considered to be very challenging are now easier to be thought of with the results and the techniques presented here. For example the classical spin-photon interface presented in chapter 6 can be converted into a quantum interface by converting the red and the blue photons into photons with identical frequencies but with orthogonal polarizations, using birefringent elements and optical modulators [63]. The technique of chapter 7, to create a precise amount of nuclear spin polarization with a pump laser and to probe with a second weak laser, can be used to measure whether or how much the nuclear spin polarization changes the linewidth of the QD transitions and the width of the nuclear spin distribution. The latter is predicted to decrease after dragging [17]. Some other experiments are listed and explained in more detail below:

8.1. Distant spin-spin entanglement

The realization of a spin-photon interface constitutes a key step toward implementation of quantum information processing protocols such as non-deterministic spin entanglement between distant spins. Even though elimination of laser background through polarization suppression in our scheme results in the loss of correlations between the spin-state and the emitted photon polarization, the fact that the photons emitted by the two spin-states have different energies ensures that by driving both the red and the blue trion transitions resonantly with a π-pulse, we could generate the entangled spin-photon state $|\psi\rangle = (|\uparrow, 1_{blue}\rangle - |\downarrow, 1_{red}\rangle)/\sqrt{2}$, starting from an initial electron spin in state $(|\uparrow\rangle - |\downarrow\rangle)/\sqrt{2}$. As was demonstrated by Monroe and co-workers [64] for two trapped ions, such entangled spin-photon states generated from two distant QDs could be used to achieve spin entanglement conditioned upon coincidence detection of one blue-trion and one red-trion photon at the output of a Hong-Ou-Mandel interferometer. We emphasize that in the non-deterministic entanglement experiments using trapped ions, the entanglement fidelity was primarily limited by the detector dark counts that were $\sim 20\%$ of the signal photons: these considerations highlight the significance of the factor of 20 improvement in the fluorescence to background photon ratio we have achieved in our experiments, as compared to prior work [1]. The principal challenge in realization of the distant QD spin-entanglement scheme is the identification of two QDs with similar enough trion resonances. Even if the two QD trion resonance energies are not exactly identical, it is possible to generate identical QD photons by off-resonant Rayleigh scattering In the case of two non-identical QDs, it would be advantageous to use the Voigt geometry, where one starts from an initial electron spin state $|\uparrow_x\rangle$ and upon scattering of a laser photon project the system onto the state $|\phi\rangle = (|\uparrow_x, 1_{blue}\rangle + |\downarrow_x, 1_{red}\rangle)/\sqrt{2}$. In this case, the energy difference between the blue and the red photons is determined exclusively by the electron Zeeman energy. On the other hand, nearly identical QD pairs that can be tuned onto resonance using the gate-voltage-induced dc-Stark shift have already been identified in other experiments [65]. More importantly, demonstration of two-photon interference of single-photon pulses generated by two different QDs has also been demonstrated [66]. Locking of trion resonances to the resonant laser field via dynamical nuclear spin polarization could be used to ensure that the electron Zeeman splitting of the two QDs are rendered identical [17].

A particularly exciting possible extension for the spin-photon interface demonstrated here is for a double-QD system consisting of a gate-defined and a self-assembled QD [67]. It has been shown that in a coupled-QD system, optical transitions in a neutral QD could be used to monitor the spin-state of a single-electron charged QD [68]. If realized, a double QD spin-photon interface along the lines we describe in this chapter could be used to generate distant spin entanglement between two gated QDs. Another interesting extension is the implementation of our method in a single-hole charged QD where the positively-charged trion transition selection rules are identical to the ones we considered [2]; because of the larger T_2^*-time of the hole spin, distant hole-spin entanglement protocol would be easier to verify experimentally.

8.2. Quantum nondemolition (QND) measurement

From a quantum measurement perspective, our electron spin measurement experiments in chapter 6 realize a positive operator valued measure (POVM) with measurement operators $\hat{E}_1 = p_1 |\uparrow\rangle\langle\uparrow|$ and $\hat{E}_2 = |\downarrow\rangle\langle\downarrow| + (1 - p_1)|\uparrow\rangle\langle\uparrow|$ [69]. For the parameters of fig 6.2, the probability $p_1 \simeq 0.02$ for a measurement time of $\sim 0.5\mu s$, but can be increased to 0.1 as mentioned before. If the collection efficiency and the excitation laser power were increased such that $p_1 \sim 1.0$ was achieved, our scheme would constitute a single-shot QND measurement of the electron spin which is crucial for fault-tolerant quantum computation. We estimate that such a spin measurement [52] is within reach using our scheme. Collection efficiencies from microcavities have been predicted to be as high as 35% [70]. One possibility is to use QDs coupled to gated photonic crystal microcavities that are engineered for high collection efficiency [71, 72]. We predict that such structures could give up to a factor of 15 improvement in the overall collection efficiency. In addition, given the observation that a sizable fraction of QDs have vanishing in-plane heavy-hole g-factors [46], it is plausible to expect much smaller spin-flip spontaneous emission rate from the trion state γ, as compared to what we observed in our experiments. A smaller branching ratio will reduce the laser back action allowing a non-demolition measurement to be performed on a longer time scale, and hence with better efficiency. By combining the above two factors, achieving a factor of more than 20 improvement in the measurement efficiency p_1 necessary for a single-shot QND measurement appears feasible. We note that recently an optical single shot electron spin measurement has been realized in a double quantum dot structure where the transitions of one dot give information on the electron spin residing in the other dot [73].

8.3. Electron spin resonance

The background-free spin-photon interface that we have realized could be considered as a non-deterministic method for conditional spin-state initialization with high fidelity. The deterministic methods for spin-state preparation rely on optical pumping where a fidelity exceeding 99% could only be achieved on a timescale $\sim 10\mu s$ for Faraday [11] and $\sim 10ns$ for Voigt geometry [74]. In contrast, detection of a single resonance fluorescence photon in our case prepares the spin state with the same level of initialization fidelity on a timescale limited only by the detector response time [2]. For the APD's we have used, this timescale is 300ps; however with faster single-photon detectors, it would be straightforward to achieve a timescale of 40 ps. Given the ultra-short $T_2^* \sim 2nsec$ characteristic of electron spins in self-assembled QDs [16], fast spin-state initialization is important for protocols relying on preparation of the spin in a superposition state. We also predict that fast spin initialization could be useful in carrying out conditional electron spin resonance (ESR) measurements [53] without the need for a pulsed microwave source; we envision here that the detection of a photon initializes the electron spin in the spin-up state, which then undergoes Rabi oscillations under the influence of the continuous-wave resonant microwave field. The likelihood of detecting a second photon at a time τ later will oscillate with this Rabi coupling, such that photon coincidences at time delay τ will reveal information about coherent spin rotation.

8.4. Measurement of in-plane hole hyperfine interaction

We measured the hole hyperfine interaction along the z-direction (sample growth axis). Interaction along this direction is relevant for decoherence if the basis of the hole spin qubit is also along the z-direction. In the Voigt geometry the lateral hyperfine interaction with the hole is of relevance. The interaction strength in the lateral direction can be measured if the presented experiments of chapter 7 are repeated in the Voigt geometry. It is predicted that this interaction is due to hh-lh mixing and it's strength is expected to be $\sim 1\%$ of the Ising-type interaction [55, 56]. Measurement of this interaction would hence require ~ 10 times better measurement accuracy.

8.5. Study of heavy-hole light-hole mixing

Heavy hole(hh)-light hole(lh) mixing has been discussed in section 5.3. The mixing results in a weak diagonal decay channel with the branching ratio $|\epsilon_+|^2 + |\epsilon_-|^2 = \eta = \frac{\gamma}{\Gamma}$, assuming equal rates for electon-hh and electron-lh recombinations. As discussed in section 5.3 we measured the same η of $\frac{1}{250}$ for 10 different dots, raising the question on the effect of a finite sample tilt with respect to the magnetic field. To eliminate this question and to obtain a reliable value for $|\epsilon_+|^2 + |\epsilon_-|^2$ the sample tilt can be

---

[2]We emphasize that the average waiting time for achieving this single-photon-detection-based initialization is still $\sim 10\mu s$ for the Faraday geometry that we used.

controlled by mounting the sample on two perpendicularly oriented goniometers. The Hall current from a Hall bar placed next to the sample would enable to follow the goniometers.

After measuring $|\epsilon_+|^2+|\epsilon_-|^2$ the next question would be the ratio $\frac{|\epsilon_+|}{|\epsilon_-|}$. Comparing the size of the small peak at $-8\,GHz$ in Fig. 5.5 with the large peak would enable to calculate the ratio $\frac{|\epsilon_+|}{|\epsilon_-|}$. The signal for the large peak originates from the path $|\downarrow\rangle \xrightarrow{\frac{\Omega^2}{\Gamma}} |\Downarrow\uparrow\downarrow\rangle \xrightarrow{\Gamma_{sp}} |\downarrow\rangle$. The signal for the small peak originates from the path $|\uparrow\rangle \xrightarrow{\frac{(\Omega|\epsilon_-|)^2}{\Gamma}} |\Downarrow\uparrow\downarrow\rangle \xrightarrow{\Gamma_{sp}} |\downarrow\rangle$.

Here Ω is the Rabi frequency due to the σ^- polarized component of the linearly polarized laser. Only the σ^- component of the laser can drive the diagonal transition and can do so due to the $|\Downarrow\rangle_{th}$ mixing. The pump rate for the diagonal transition is reduced by the mixing factor $|\epsilon_-|^2$. The ratio of the peak sizes is equal to the ratio of the trion populations which is in turn proportional to the pump rates so that this ratio is equal to $|\epsilon_-|^2$.

A. Bibliography

[1] C.-Y. Lu, Y. Zhao, A. N. Vamivakas, C. Matthiesen, S. Fält, A. Badolato, and M. Atatüre, "Direct measurement of spin dynamics in inas/gaas quantum dots using time-resolved resonance fluorescence," *Phys. Rev. B*, vol. 81, p. 035332, Jan 2010. c, 9, 28, 43

[2] B. D. Gerardot, D. Brunner, P. A. Dalgarno, P. Ohberg, S. Seidl, M. Kroner, K. Karrai, N. G. Stoltz, P. M. Petroff, and R. J. Warburton, "Optical pumping of a single hole spin in a quantum dot," *Nature*, vol. 451, pp. 441–444, 2008. c, 33, 44

[3] D. Heiss, S. Schaeck, H. Huebl, M. Bichler, G. Abstreiter, J. J. Finley, D. V. Bulaev, and D. Loss, "Observation of extremely slow hole spin relaxation in self-assembled quantum dots," *Phys. Rev. B*, vol. 76, p. 241306, Dec 2007. 33

[4] M. Kroutvar, Y. Ducommun, D. Heiss, M. Bichler, D. Schuh, G. Abstreiter, and J. J. Finley, "Optically programmable electron spin memory using semiconductor quantum dots," *Nature*, vol. 432, pp. 81–84, 2004. c

[5] J. R. Petta, A. C. Johnson, J. M. Taylor, E. A. Laird, A. Yacoby, M. D. Lukin, C. M. Marcus, M. P. Hanson, and A. C. Gossard, "Coherent Manipulation of Coupled Electron Spins in Semiconductor Quantum Dots," *Science*, vol. 309, no. 5744, pp. 2180–2184, 2005. c, 33

[6] M. H. Mikkelsen, J. Berezovsky, N. G. Stoltz, L. A. Coldren, and D. D. Awschalom, "Optically detected coherent spin dynamics of a single electron in a quantum dot," *Nat Phys*, vol. 3, pp. 770–773, 2007. 33

[7] D. Brunner, B. D. Gerardot, P. A. Dalgarno, G. Wst, K. Karrai, N. G. Stoltz, P. M. Petroff, and R. J. Warburton, "A Coherent Single-Hole Spin in a Semiconductor," *Science*, vol. 325, no. 5936, pp. 70–72, 2009. 33

[8] A. Greilich, D. R. Yakovlev, A. Shabaev, A. L. Efros, I. A. Yugova, R. Oulton, V. Stavarache, D. Reuter, A. Wieck, and M. Bayer, "Mode Locking of Electron Spin Coherences in Singly Charged Quantum Dots," *Science*, vol. 313, no. 5785, pp. 341–345, 2006. c

[9] A. Imamoglu, D. D. Awschalom, G. Burkard, D. P. DiVincenzo, D. Loss, M. Sherwin, and A. Small, "Quantum information processing using quantum dot spins and cavity qed," *Phys. Rev. Lett.*, vol. 83, pp. 4204–4207, Nov 1999. c, 25

[10] R. Hanson and D. D. Awschalom, "Coherent manipulation of single spins in semiconductors," *Nature*, vol. 453, pp. 1043–1049, 2008. c, 33

[11] M. Atatüre, J. Dreiser, A. Badolato, A. Högele, K. Karrai, and A. Imamoglu, "Quantum-Dot Spin-State Preparation with Near-Unity Fidelity," *Science*, vol. 312, no. 5773, pp. 551–553, 2006. c, 17, 26, 35, 45

[12] J. Berezovsky, M. H. Mikkelsen, N. G. Stoltz, L. A. Coldren, and D. D. Awschalom, "Picosecond Coherent Optical Manipulation of a Single Electron Spin in a Quantum Dot," *Science*, vol. 320, no. 5874, pp. 349–352, 2008. c

[13] D. Press, T. D. Ladd, B. Zhang, and Y. Yamamoto, "Complete quantum control of a single quantum dot spin using ultrafast optical pulses," *Nature*, vol. 456, pp. 218–221, 2008. c

[14] S. T. Yilmaz, P. Fallahi, and A. Imamoglu, "Quantum-dot-spin single-photon interface," *Phys. Rev. Lett.*, vol. 105, p. 033601, Jul 2010. c, 25, 34, 35

[15] A. N. Vamivakas, C.-Y. Lu, C. Matthiesen, Y. Zhao, S. Falt, A. Badolato, and M. Atature, "Observation of spin-dependent quantum jumps via quantum dot resonance fluorescence," *Nature*, vol. 467, pp. 297–300, 2010. c

[16] J. Dreiser, M. Atatüre, C. Galland, T. Müller, A. Badolato, and A. Imamoglu, "Optical investigations of quantum dot spin dynamics as a function of external electric and magnetic fields," *Phys. Rev. B*, vol. 77, p. 075317, Feb 2008. c, 10, 11, 35, 45

[17] C. Latta, A. Hogele, Y. Zhao, A. N. Vamivakas, P. Maletinsky, M. Kroner, J. Dreiser, I. Carusotto, A. Badolato, D. Schuh, W. Wegscheider, M. Atature, and A. Imamoglu, "Confluence of resonant laser excitation and bidirectional quantum-dot nuclear-spin polarization," *Nat Phys*, vol. 5, pp. 758–763, 2009. c, 26, 33, 34, 35, 41, 43, 44

[18] P. Fallahi, S. T. Yilmaz, and A. Imamoglu, "Measurement of a heavy-hole hyperfine interaction in ingaas quantum dots using resonance fluorescence," *Phys. Rev. Lett.*, vol. 105, p. 257402, Dec 2010. c, 33

[19] C. W. Lai, P. Maletinsky, A. Badolato, and A. Imamoglu, "Knight-field-enabled nuclear spin polarization in single quantum dots," *Phys. Rev. Lett.*, vol. 96, p. 167403, Apr 2006.

[20] P. Maletinsky, A. Badolato, and A. Imamoglu, "Dynamics of quantum dot nuclear spin polarization controlled by a single electron," *Phys. Rev. Lett.*, vol. 99, p. 056804, Aug 2007. c

[21] H. J. Kimble, "The quantum internet," *Nature*, vol. 453, pp. 1023–1030, 2008. c

[22] L. Childress, J. M. Taylor, A. S. Sørensen, and M. D. Lukin, "Fault-tolerant quantum communication based on solid-state photon emitters," *Phys. Rev. Lett.*, vol. 96, p. 070504, Feb 2006. c

[23] A. Högele, *Laser spectroscopy of single charge-tunable quantum dots*. PhD thesis, LMU Munich, 2005. 1, 14

[24] J. Dreiser, *Optical study, preparation and measurement of a single quantum-dot spin*. PhD thesis, ETH Zurich, 2007.

[25] M. Kroner, *Resonant photon-exciton interaction in semiconductor quantum dots*. PhD thesis, LMU Munich, 2008.

[26] M. Atatüre, "Quantum optics with quantum dot spins (habilitation thesis)," 2007. 1

[27] S. Seidl, M. Kroner, P. A. Dalgarno, A. Högele, J. M. Smith, M. Ediger, B. D. Gerardot, J. M. Garcia, P. M. Petroff, K. Karrai, and R. J. Warburton, "Absorption and photoluminescence spectroscopy on a single self-assembled charge-tunable quantum dot," *Phys. Rev. B*, vol. 72, p. 195339, Nov 2005. 1

[28] R. J. Warburton, B. T. Miller, C. S. Dürr, C. Bödefeld, K. Karrai, J. P. Kotthaus, G. Medeiros-Ribeiro, P. M. Petroff, and S. Huant, "Coulomb interactions in small charge-tunable quantum dots: A simple model," *Phys. Rev. B*, vol. 58, pp. 16221–16231, Dec 1998. 1, 2

[29] R. J. Warburton, C. Schaflein, D. Haft, F. Bickel, A. Lorke, K. Karrai, J. M. Garcia, W. Schoenfeld, and P. M. Petroff, "Optical emission from a charge-tunable quantum ring," *Nature*, vol. 405, pp. 926–929, 2000. 1

[30] S. Raymond, S. Fafard, P. J. Poole, A. Wojs, P. Hawrylak, S. Charbonneau, D. Leonard, R. Leon, P. M. Petroff, and J. L. Merz, "State filling and time-resolved photoluminescence of excited states in $in_xga_{1-x}as$/gaas self-assembled quantum dots," *Phys. Rev. B*, vol. 54, pp. 11548–11554, Oct 1996. 2

[31] R. Heitz, M. Veit, N. N. Ledentsov, A. Hoffmann, D. Bimberg, V. M. Ustinov, P. S. Kop'ev, and Z. I. Alferov, "Energy relaxation by multiphonon processes in inas/gaas quantum dots," *Phys. Rev. B*, vol. 56, pp. 10435–10445, Oct 1997. 2

[32] R. Heitz, A. Kalburge, Q. Xie, M. Grundmann, P. Chen, A. Hoffmann, A. Madhukar, and D. Bimberg, "Excited states and energy relaxation in stacked inas/gaas quantum dots," *Phys. Rev. B*, vol. 57, pp. 9050–9060, Apr 1998. 2

[33] R. J. Warburton, C. Schulhauser, D. Haft, C. Schäflein, K. Karrai, J. M. Garcia, W. Schoenfeld, and P. M. Petroff, "Giant permanent dipole moments of excitons in semiconductor nanostructures," *Phys. Rev. B*, vol. 65, p. 113303, Feb 2002. 3

[34] C. Schulhauser, D. Haft, R. J. Warburton, K. Karrai, A. O. Govorov, A. V. Kalameitsev, A. Chaplik, W. Schoenfeld, J. M. Garcia, and P. M. Petroff, "Magneto-optical properties of charged excitons in quantum dots," *Phys. Rev. B*, vol. 66, p. 193303, Nov 2002. 3

[35] C. Schulhauser, D. Haft, R. J. Warburton, K. Karrai, A. O. Govorov, A. V. Kalameitsev, A. Chaplik, W. Schoenfeld, J. M. Garcia, and P. M. Petroff, "Magneto-optical properties of charged excitons in quantum dots," *Phys. Rev. B*, vol. 66, p. 193303, Nov 2002. 4

[36] C. Schulhauser, *Electronische Quantenpunktzustände induziert durch Photonemission*. PhD thesis, LMU Munich, 2004. 4

[37] V. Zwiller and G. Bjork, "Improved light extraction from emitters in high refractive index materials using solid immersion lenses," *Journal of Applied Physics*, vol. 92, no. 2, pp. 660–665, 2002. 7

[38] Y. Yamamoto and A. Imamoglu, *Mesoscopic Quantum Optics*. Wiley, New York, 1999. 10

[39] D. V. Averin and Y. V. Nazarov, "Virtual electron diffusion during quantum tunneling of the electric charge," *Phys. Rev. Lett.*, vol. 65, pp. 2446–2449, Nov 1990. 11

[40] R. Loudon, *The quantum theory of light*. Oxford University Press, 2000. 11, 30

[41] X. Zeng, C. Liang, and Y. An, "Far-Field Propagation of an Off-Axis Gaussian Wave," *Appl. Opt.*, vol. 38, pp. 6253–6256, 1999. 12

[42] L. G. Guoy *Comp. Rend. Acad. Sci. Paris*, vol. 110, pp. 1251–1253, 1890. 12

[43] M. Kroner, S. Remi, and A. H "Resonant saturation laser spectroscopy of a single self-assembled quantum dot," *Physica E: Low-dimensional Systems and Nanostructures*, vol. 40, no. 6, pp. 1994 – 1996, 2008. 13th International Conference on Modulated Semiconductor Structures. 14, 19

[44] M. Kroner, K. M. Weiss, B. Biedermann, S. Seidl, A. W. Holleitner, A. Badolato, P. M. Petroff, P. Öhberg, R. J. Warburton, and K. Karrai, "Resonant two-color high-resolution spectroscopy of a negatively charged exciton in a self-assembled quantum dot," *Phys. Rev. B*, vol. 78, p. 075429, Aug 2008. 16

[45] W. Sheng and P. Hawrylak, "Spin polarization in self-assembled quantum dots," *Phys. Rev. B*, vol. 73, p. 125331, Mar 2006. 17

[46] G. Fernandez, T. Volz, R. Desbuquois, A. Badolato, and A. Imamoglu, "Optically tunable spontaneous raman fluorescence from a single self-assembled ingaas quantum dot," *Phys. Rev. Lett.*, vol. 103, p. 087406, Aug 2009. 16, 17, 44

[47] M. Atatüre, J. Dreiser, A. Badolato, and A. Imamoglu, "Observation of faraday rotation from a single confined spin," *Nat Phys*, vol. 3, pp. 101–106, 2007. 22

[48] B. B. Blinov, D. L. Moehring, L.-M. Duan, and C. Monroe, "Observation of entanglement between a single trapped atom and a single photon," *Nature*, vol. 428, pp. 153–157, 2004. 25

[49] E. Togan, Y. Chu, A. S. Trifonov, L. Jiang, J. Maze, L. Childress, M. V. G. Dutt, A. S. Sorensen, P. R. Hemmer, A. S. Zibrov, and M. D. Lukin, "Quantum entanglement between an optical photon and a solid-state spin qubit," *Nature*, vol. 466, pp. 730–734, 2004. 25

[50] A. Nick Vamivakas, Y. Zhao, and M. Lu, Chao-Yang and Atature, "Spin-resolved quantum-dot resonance fluorescence," *Nat Phys*, vol. 5, pp. 198–202, 2009. 25, 34

[51] E. B. Flagg, A. Muller, J. W. Robertson, S. Founta, D. G. Deppe, M. Xiao, W. Ma, G. J. Salamo, and C. K. Shih, "Resonantly driven coherent oscillations in a solid-state quantum emitter," *Nat Phys*, vol. 5, pp. 203–207, 2009. 25, 34

[52] R. Hanson, L. P. Kouwenhoven, J. R. Petta, S. Tarucha, and L. M. K. Vandersypen, "Spins in few-electron quantum dots," *Rev. Mod. Phys.*, vol. 79, pp. 1217–1265, Oct 2007. 33, 44

[53] F. H. L. Koppens, C. Buizert, K. J. Tielrooij, I. T. Vink, K. C. Nowack, T. Meunier, L. P. Kouwenhoven, and L. M. K. Vandersypen, "Driven coherent oscillations of a single electron spin in a quantum dot," *Nature*, vol. 442, pp. 766–771, 2006. 45

[54] W. A. Coish and D. Loss, "Hyperfine interaction in a quantum dot: Non-markovian electron spin dynamics," *Phys. Rev. B*, vol. 70, p. 195340, Nov 2004. 33

[55] J. Fischer, W. A. Coish, D. V. Bulaev, and D. Loss, "Spin decoherence of a heavy hole coupled to nuclear spins in a quantum dot," *Phys. Rev. B*, vol. 78, p. 155329, Oct 2008. 33, 41, 45

[56] C. Testelin, F. Bernardot, B. Eble, and M. Chamarro, "Hole–spin dephasing time associated with hyperfine interaction in quantum dots," *Phys. Rev. B*, vol. 79, p. 195440, May 2009. 33, 45

[57] H. Kurtze, D. R. Yakovlev, D. Reuter, A. D. Wieck, and M. Bayer, "Hyperfine interaction of electron and hole spin with the nuclei in (In,Ga)As/GaAs quantum dots," *ArXiv e-prints*, May 2009. 33

[58] P. Desfonds, B. Eble, F. Fras, C. Testelin, F. Bernardot, M. Chamarro, B. Urbaszek, T. Amand, X. Marie, J. M. Gerard, V. Thierry-Mieg, A. Miard, and A. Lemaitre, "Electron and hole spin cooling efficiency in inas quantum dots: The role of nuclear field," *Applied Physics Letters*, vol. 96, no. 17, p. 172108, 2010. 33

[59] M. Kroner, K. M. Weiss, B. Biedermann, S. Seidl, S. Manus, A. W. Holleitner, A. Badolato, P. M. Petroff, B. D. Gerardot, R. J. Warburton, and K. Karrai, "Optical detection of single-electron spin resonance in a quantum dot," *Phys. Rev. Lett.*, vol. 100, p. 156803, Apr 2008. 36

[60] W. Yang and L. J. Sham, "Intrinsic bidirectional dynamic nuclear polarization by optically pumped trions in quantum dots," *ArXiv e-prints*, Mar. 2010. 41

[61] X. Xu, W. Yao, B. Sun, D. G. Steel, A. S. Bracker, D. Gammon, and L. J. Sham, "Optically controlled locking of the nuclear field via coherent dark-state spectroscopy," *Nature*, vol. 459, pp. 1105–1109, 2009. 41

[62] E. A. Chekhovich, A. B. Krysa, M. S. Skolnick, and A. I. Tartakovskii, "Direct measurement of the hole-nuclear spin interaction in single inp/gainp quantum dots using photoluminescence spectroscopy," *Phys. Rev. Lett.*, vol. 106, p. 027402, Jan 2011. 41

[63] W. Gao. personal communication, 2011. 43

[64] D. L. Moehring, P. Maunz, S. Olmschenk, K. C. Younge, D. N. Matsukevich, L.-M. Duan, and C. Monroe, "Entanglement of single-atom quantum bits at a distance," *Nature*, vol. 449, pp. 68–71, 2007. 43

[65] A. Laucht, J. M. Villas-Bôas, S. Stobbe, N. Hauke, F. Hofbauer, G. Böhm, P. Lodahl, M.-C. Amann, M. Kaniber, and J. J. Finley, "Mutual coupling of two semiconductor quantum dots via an optical nanocavity," *Phys. Rev. B*, vol. 82, p. 075305, Aug 2010. 44

[66] E. B. Flagg, A. Muller, S. V. Polyakov, A. Ling, A. Migdall, and G. S. Solomon, "Interference of single photons from two separate semiconductor quantum dots," *Phys. Rev. Lett.*, vol. 104, p. 137401, Apr 2010. 44

[67] H. Engel, J. M. Taylor, M. D. Lukin, and A. Imamoglu, "Quantum optical interface for gate-controlled spintronic devices," *ArXiv Condensed Matter e-prints*, Dec. 2006. 44

[68] D. Kim, S. E. Economou, i. m. c. C. Bădescu, M. Scheibner, A. S. Bracker, M. Bashkansky, T. L. Reinecke, and D. Gammon, "Optical spin initialization and nondestructive measurement in a quantum dot molecule," *Phys. Rev. Lett.*, vol. 101, p. 236804, Dec 2008. 44

[69] M. A. Nielsen and I. L. Chuang, *Quantum Computation and Quantum Information*. Cambridge University Press; 1 edition, 2000. 44

[70] M. Larqu, T. Karle, I. Robert-Philip, and A. Beveratos, "Optimizing h1 cavities for the generation of entangled photon pairs," *New Journal of Physics*, vol. 11, no. 3, p. 033022, 2009. 44

[71] M. T. Rakher, N. G. Stoltz, L. A. Coldren, P. M. Petroff, and D. Bouwmeester, "Externally mode-matched cavity quantum electrodynamics with charge-tunable quantum dots," *Phys. Rev. Lett.*, vol. 102, p. 097403, Mar 2009. 44

[72] D. Englund, A. Faraon, A. Majumdar, N. Stoltz, P. Petroff, and J. Vuckovic 44

[73] A. N. Vamivakas, C.-Y. Lu, C. Matthiesen, Y. Zhao, S. Falt, A. Badolato, and M. Atature, "Observation of spin-dependent quantum jumps via quantum dot resonance fluorescence," *Nature*, vol. 467, pp. 297–300, 2010. 44

[74] X. Xu, Y. Wu, B. Sun, Q. Huang, J. Cheng, D. G. Steel, A. S. Bracker, D. Gammon, C. Emary, and L. J. Sham, "Fast spin state initialization in a singly charged inas-gaas quantum dot by optical cooling," *Phys. Rev. Lett.*, vol. 99, p. 097401, Aug 2007. 45

Acknowledgment

PhD requires dedication. You get obsessed with the measurements, you lose your connection with the normal social life. But the daily routine in the lab itself is a form of social life and I think it is very rich. There is science, responsibility, friendship, patience, altruism and competition, glory and desperation, pragmatism and the desire to understand ... So many things depend on just what you will decide in situations with many parameters. All these things create the perfect conditions to know who you are. In the last 4.5 years my life was almost solely defined by the PhD experience. I have been through a transformation and thanks to a bunch of people it was a good transformation. I am sincerely and deeply thankful to them. Atac's influential role in my life is unquestionable. Besides being a role model as a scientist and as a human-being, I feel that he did so many things for me just out of good will. It was his confidence in me that made me decide to do a PhD in physics in the first place.

Also, I am thankful to him that he created a group with so many nice people and maintained it as such. I am thankful to Mete and Jan from my earlier days in the group. They patiently taught me what they had learned over the years by experience. They provided a solid background upon which I could build. Thanks to Thomas and Gemma for the year we spent together in lab, which then developed into real friendships. Thanks to Dora, she really had to be tolerant with me during my last days in the lab, a time when I was very stressful. Special thanks to my lab partner Parisa. We have been through so many things but it was a very happy ending in the end. Thanks to Ajit for reminding me, regardless of all the details, why we are doing science. G-foor is a place where I feel most at home. Regardless of how bad and isolated I felt, coming to G-floor and greeting someone always made me feel better, more confident, more optimistic. I am thankful to all the people in the group, past and present, for creating this supportive atmosphere. Even though there are people who have been closer to me than the others it is difficult for me to give names. Whenever I think to leave out someone's name, a nice experience with this person comes to my mind. In Zurich I moved many times and I had the opportunity of having great flat mates like Onur, Hakan and Emre. The presence of my old friend Can in Switzerland probably slowed down my progress, but thanks to him I am graduating more happily. I am thankful to all the friends outside of campus, both in Switzerland and the ones from Turkey with whom we are still in touch. I am thankful to my family for their support of my decisions, even though it meant and it will continue to mean separation for them.

List of Figures

4.1. Sketch of the sample structure (left) together with the band diagram along the growth direction (right). V_g^0 is the built-in voltage due to Shottky barrier and V_g is the applied voltage. 2

4.2. μ-photoluminescence spectrum as a function of gate voltage. 3

4.3. Magnetic field dependence of the X^- resonances for the electron spin-up and spin-down transitions. 4

4.4. Sketch of the confocal microscope immersed into the liquid helium bath cryostat. 5

4.5. The path of the light ray from a beam focused on the sample. 7

5.1. Level diagram describing the singly charged QD in the presence of an external magnetic field in the z-direction. For clarity the transitions involving the level $|\downarrow\uparrow\Downarrow\rangle$ are not shown. Shown are the coupling of the blue vertical transition to the σ^+-polarized light field with Rabi frequency Ω_R, the spontaneous emission rate Γ, the weak diagonal decay rate $\gamma = \eta\Gamma$ with $\eta << 1$ being the branching ratio, the hyperfine interaction induced electron spin state mixing with the strength $\hbar\Omega_N$ and the electron spin flip rate κ as a result of the cotunneling processes. 10

5.2. Geometry of the QD excitation. A linearly polarized Gaussian laser beam is focused on the QD considered as a dipole along the x-direction. A detector in the far field measures the composite field over the solid angle bounded by θ^{max}. 11

5.3. Resonance fluorescence (RF) counts as a function of gate voltage at $B = 0\,T$ when the QD is excited by a laser with power $P_L^{det} \approx P_{sat}$. The curve is fitted to a Lorentzian (top). The contrast P_{scat}^{det}/P_L^{det} and FWHM of the linewidth $\Delta\omega$ as a function of detector photo current. The data is fitted to 5.34 and 5.33 given in the text (bottom). 15

5.4. (a) Resonance fluorescence data as a function of laser detuning $\frac{\Delta\omega}{2\pi}$ and gate voltage. At $B = 500\,mT$ a linearly polarized laser is scanned across the QD transitions and the gate voltage is stepped within the single electron regime. The signals from the transitions $|\uparrow\rangle \leftrightarrow |\downarrow\uparrow\Uparrow\rangle$ (above $\frac{\Delta\omega}{2\pi} = 0 GHz$) and $|\downarrow\rangle \leftrightarrow |\downarrow\uparrow\Downarrow\rangle$ (below $\frac{\Delta\omega}{2\pi} = 0 GHz$) are visible in some gate voltage ranges. Since, apart at the edges of the gate voltage plateaus, the cotunneling rate κ is small, the signals disappear due to electron spin pumping (left). A second linearly polarized laser is introduced which at a gate voltage of $270\,mV$ becomes on resonance with the $|\downarrow\rangle \leftrightarrow |\downarrow\uparrow\Downarrow\rangle$ transition. When the first laser hits the resonance $|\uparrow\rangle \leftrightarrow |\downarrow\uparrow\Uparrow\rangle$ at $\frac{\Delta\omega}{2\pi} \approx 7\,GHz$ ($|\uparrow\rangle \leftrightarrow |\downarrow\uparrow\Downarrow\rangle$ at $\frac{\Delta\omega}{2\pi} \approx -2.5\,GHz$) the double resonance condition leads to the recovery (partial recovery) of the signal. 16

5.5. Cross section of the left panel of 5.4 showing a laser scan at a gate voltage of $225\,mV$ (left) and it's zoom-in (right). The large peak is the resonance with $|\downarrow\rangle \leftrightarrow |\downarrow\uparrow\Downarrow\rangle$ transition and the small signal at $\frac{\Delta\omega}{2\pi} = -9\,GHz$ is the resonance with the diagonal transition $|\uparrow\rangle \leftrightarrow |\downarrow\uparrow\Downarrow\rangle$. 18

5.6. Poincare sphere showing the arbitrary measurement basis state $|\psi^+\rangle$. 21

5.7. (a) Gate voltage scan at $B = 1\,T$ showing the dispersive signal from the $|\downarrow\rangle \leftrightarrow |\downarrow\uparrow\Downarrow\rangle$ transition. The linearly polarized excitation laser makes an angle $\theta_L = 0.01$ to $|x\rangle$. The reflected laser photons together with the dot photons are sent to an APD after passing through a polarizer oriented along $|y\rangle$. 22

6.1. (a) Energy level diagram for a quantum dot (QD) charged with a single electron. (b)&(c) Differential Transmission (DT) signal as a function of gate voltage and laser frequency at $B = 1T$ and $B = 0T$. At $B = 0T$ DT signal is seen at gate voltages where the QD is singly charged. At $B = 1T$ the DT signal (white points) in the middle of the plateau disappears due to spin pumping. Dashed line corresponds to the gate voltage trace in Fig. 6.2(a) 26

List of Figures   XI

6.2. (a) Resonance fluorescence (RF) signal from the QD as the gate voltage is scanned along the red dashed line on Fig. 6.1b. Laser power is well below the QD saturation, $P = 0.1 \cdot P_{sat}$. The gray trace shows the measurement background obtained at the same laser power with the laser frequency fully detuned from the QD transition. On resonance the ratio of RF photons to laser background exceeds 200. (b) A typical time trace recorded from the avalanche photo-diode (APD) with a 200 nsec time resolution with a resonant laser with $P = 0.1 \cdot P_{sat}$. Each pulse arises from the detection of a single photon, which indicates that the spin is in the $|\uparrow\rangle$ state with $(99.2 \pm 0.1)\%$ fidelity. (c) G_2 curve obtained by measuring photon coincidences on two APDs on nsec timescales. The expected antibunching behavior for a single emitter is observed, with a spontaneous emission rate $\Gamma \sim 10^9 s^{-1}$. The G_2 curve does not reach zero at $\tau = 0$ due to the finite time resolution $\sim 450\,psec$ of the Hanbury-Brown and Twiss measurement set-up. (d) Unnormalized photon correlation, G_2 curve obtained from $\sim 60,000$ traces such as the trace in (b) for $P = 0.1 \cdot P_{sat}$. Solid line is an exponential fit with a decay time $\tau_{decay} = (540 \pm 40)$nsec, corresponding to the cotunneling limited lifetime of the $|\uparrow\rangle$ state. 27

6.3. (a) G_2 (red) and nshot (black) measurements obtained at a gate voltage in the middle of the plateau and laser intensity $P = P_{sat}$. Solid lines are exponential fits with decay times $(840 \pm 40)ns$ and $(860 \pm 20)ns$ respectively, demonstrating the agreement between the two measurement methods. (b)&(c) G_2 curves obtained from a second quantum dot with a laser power $P = 0.1 \cdot P_{sat}$ and at two gate voltages close to the plateau edge (large cotunneling) and $7.5mV$ apart. The solid lines are exponential fits, showing decay times of $(1.0 \pm 0.1)\mu s$ and $(2.5 \pm 0.2)\mu s$ respectively. These decay times are direct measurements of cotunneling-limited spin decay rates at the given gate voltages, with the faster decay, (b), corresponding to the gate voltage closer to the plateau edge as expected. 29

7.1. (a) Resonance fluorescence (RF) signal from the blue trion transition vs gate voltage and pump laser detuning, Δ at $B = 4T$ and $P = P_{sat}/2$. Remainder of the experiments are performed at the gate voltage indicated by the dashed line, where the signal is reduced ~ 4 times due to spin pumping and a large line broadening due to dynamic nuclear spin polarization is observed. Inset: energy level diagram for a quantum dot charged with a single electron. (b) Cross section of (a) across the dashed line, opposite scan directions indicated by the arrows. A total dragging range of $\sim 8GHz$ is observed. Interference with the laser background is partly responsible for the change of RF counts along the dragging range. 34

7.2. (a)&(b) Energy level diagrams showing the pump and probe lasers. The probe laser re-pumps the spin into the $|\uparrow\rangle$ state by driving the red trion (a) or the diagonal (b) transition. Dashed (solid) lines indicate the ground and excited energy levels before (after) the polarization of nuclei by the pump laser. δE_{HH} (δE_e) denotes the Overhauser shift seen by a single QD HH (electron). (c) Resonance fluorescence (RF) signal recorded as the probe laser is scanned across the diagonal (right) or the red trion (left) transitions. An arbitrary offset is added to the top two scans for clarity. Prior to the probe laser scan the pump laser is scanned across the blue trion transition and stopped at a detuning of $\Delta = 0.5 GHz$ (top) or $\Delta = -2.5 GHz$ (bottom). Solid lines are Lorentzian fits. Peak positions are shifted due to nuclear spin polarization induced by the pump laser. 35

7.3. (a) Resonance fluorescence (RF) signal as a function of probe laser frequency scanned across the diagonal transition and pump laser detuning, Δ. The position of the peak corresponding to recovered RF counts shifts due to nuclear spin polarization induced by the pump laser. (b) (green points) The position of the peaks in (a) extracted from Lorentzian fits to the data. Red points correspond to peak positions extracted from similar scans where the probe laser is scanned across the vertical transition. Dashed lines are linear fits with slopes indicated on the figure. Arrows point to the relevant axis. 37

7.4. (a) Transition energies of the red trion (red points) and the diagonal transition (green and blue points) vs pump laser detuning, Δ, at $B = 4T$. Probe laser power and slopes of linear fits (dashed lines) are indicated on the figure. The dark and light blue points are measured with the same laser power at two different gate voltages, $1mV$ apart, around the gate voltage indicated by the dashed line on Fig. 7.1(a). The dark and light green and red points are measured at two different laser powers at the gate voltage indicated by the dashed line on Fig. 7.1(a). The slopes do not change with changing probe laser power or gate voltage. Inset shows measurements at $B = 7.25T$: the axes are the same as those on the main figure. (b) Transition energies of the red trion (red points) and the diagonal transition (green points) vs Δ, measured on a second quantum dot. The vertical shift between the two green lines is due to a shift in the gate voltage, caused by fluctuations in the quantum dot charge environment. The box compares the emission wavelength, electron (ge) and heavy-hole (gh) g-factors and η for 3 different quantum dots. 38

7.5. Resonance fluorescence signal when the gate voltage is scanned in the positive direction for different laser frequencies. The magnetic field is $300\,mT$ and the excitation laser power $P \approx 10 P_{sat}$. 39

List of Figures   XIII

7.6. Shown is differential reflection signal at $300\,mT$ from successive scans in the negative direction by the weak probe laser for a second QD. In order to create nuclear spin polarization, the gate voltage and the frequency of the strong linearly polarized pump laser is held at the position corresponding to the signal indicated by the arrow in Fig. 7.5. The pump laser is blocked in the collection path. 40

Die VDM Verlagsservicegesellschaft sucht für wissenschaftliche Verlage abgeschlossene und herausragende

Dissertationen, Habilitationen, Diplomarbeiten, Master Theses, Magisterarbeiten usw.

für die kostenlose Publikation als Fachbuch.

Sie verfügen über eine Arbeit, die hohen inhaltlichen und formalen Ansprüchen genügt, und haben Interesse an einer honorarvergüteten Publikation?

Dann senden Sie bitte erste Informationen über sich und Ihre Arbeit per Email an *info@vdm-vsg.de*.

Sie erhalten kurzfristig unser Feedback!

VDM Verlagsservicegesellschaft mbH
Dudweiler Landstr. 99   Telefon +49 681 3720 174
D - 66123 Saarbrücken   Fax   +49 681 3720 1749
www.vdm-vsg.de

Die VDM Verlagsservicegesellschaft mbH vertritt

Printed by Books on Demand GmbH, Norderstedt / Germany